Defectives in the Land

Defectives in the Land

Disability and Immigration in the Age of Eugenics

DOUGLAS C. BAYNTON

The University of Chicago Press
Chicago and London

The University of Chicago Press, Chicago 60637
The University of Chicago Press, Ltd., London

Published 2016
Paperback edition 2020
Printed in the United States of America

29 28 27 26 25 24 23 22 21 20 1 2 3 4 5

ISBN-13: 978-0-226-36416-2 (cloth)
ISBN-13: 978-0-226-75863-3 (paper)
ISBN-13: 978-0-226-36433-9 (e-book)
DOI: https://doi.org/10.7208/chicago/9780226364339.001.0001

Library of Congress Cataloging-in-Publication Data
Names: Baynton, Douglas C., author.
Title: Defectives in the land : disability and immigration in the age of eugenics / Douglas C. Baynton.
Description: Chicago ; London : The University of Chicago Press, 2016. | Includes bibliographical references and index.
Identifiers: LCCN 2015036867| ISBN 9780226364162 (cloth : alk. paper) | ISBN 9780226364339 (e-book)
Subjects: LCSH: Immigrants—Medical examinations—United States. | Eugenics—United States—History. | People with disabilities—Legal status, laws, etc.—United States. | United States—Emigration and immigration—Government policy.
Classification: LCC JV6485 .B395 2016 | DDC 363.9/20973—dc23
LC record available at http://lccn.loc.gov/201503686

♾ This paper meets the requirements of ANSI/NISO Z39.48–1992 (Permanence of Paper).

For Katy and Anna

Contents

Introduction

Race and ethnicity have long been at the center of American immigration policy history, and understandably so. The formative years of federal immigration law were bracketed at one end by the Chinese Exclusion Act of 1882, and at the other by the National Origins Acts of the 1920s, which greatly reduced immigration from the "less desirable" nations of southern and eastern Europe. In the intervening years, however, a series of less controversial but historically critical laws were enacted that restricted immigration on the basis of disability, or what was commonly known at the time as "defect." During these first four decades of federal immigration law, restriction advocates, members of Congress, and Immigration Bureau officials identified defective immigrants as a dire threat to the nation. The menacing image of the defective was the principal catalyst for the rapid expansion of immigration law and the machinery of its enforcement.[1]

A great variety of disabled immigrants were refused entry, among them the deaf, blind, epileptic, and mobility impaired; people with curved spines, hernias, flat or club feet, missing limbs, and short limbs; those who were unusually short or tall; people with intellectual or psychiatric disabilities; hermaphrodites (intersexuals); men of "poor physique" and men diagnosed with "feminism"—a hormonal deficiency, relatively common at the time, that caused underdeveloped sexual organs. Others entering the country temporarily for work or pleasure were detained until some individual or organization was willing to assume legal responsibility for their oversight, support, and eventual departure. These included freak-show performers such as Juggernaut the Armless and Legless Mite, Hassan Ali the Egyptian Giant, and Miss Delphi the Orange-Headed Girl; and visiting lecturers such as Robert

Middlemiss, on a tour to educate the public on the ability of blinded war veterans to work and support themselves independently.[2]

The exclusion of individuals seen as defective was the principal objective of immigration policy for many years, but the idea of defect was also instrumental in justifying the exclusion of other types of immigrants that have received greater attention from historians.[3] Undesirable races, for example, were understood to be those in which defects proliferated. Immorality, criminality, deviant sexuality, poverty, and political radicalism were all described as manifestations of various kinds of mental defect. Infectious diseases were thought more common among those weakened by constitutional defect. And immigrants from certain nations were thought incapable of assimilating in American society, or of understanding democratic practices, owing to incapacities of mind and body that were passed down from generation to generation.[4] If, as the court in John Lennon's 1975 deportation case stated, immigration laws are "like a magic mirror, reflecting the fears and concerns of past Congresses," the early years of immigration law reflect an unusually intense fear of human defects.[5]

The many histories of immigration policy written over the past six decades have closely examined the roles played by race and ethnicity in arousing anti-immigrant sentiment and shaping legislation and enforcement. More recently, historians have investigated the important influences of gender and sexuality.[6] The extraordinary prominence of disability rhetoric in anti-immigrant literature, however, remains largely unexplored.[7] This omission is not peculiar to the history of immigration. Disability has gone largely unexplored and sometimes unremarked in many fields of history where it might be considered central—for example, in histories of industrialization, war, and the automobile, all of which have been highly efficient at producing disability at the same time that they have contributed by various means to the social marginalization of disabled people. In histories of the eugenics movement and the Holocaust, the rise of the welfare state and political opposition to it, even biographical accounts of disabled people, disability often lurks unexamined between the lines. Since the advent of the modern disability-rights movement and the subsequent emergence of disability studies in the humanities, this has begun to change, but only just, and the great majority of historians do not yet see the relevance of disability to their fields.[8]

Disability matters everywhere in history; there are no histories in which a disability analysis would be out of place and many that are diminished by its absence. In the same way that all people are defined in part by race, class, gender, and sexuality, everyone is defined in some way by disability, by its presence or ostensible absence. Moreover, the concept of disability has been

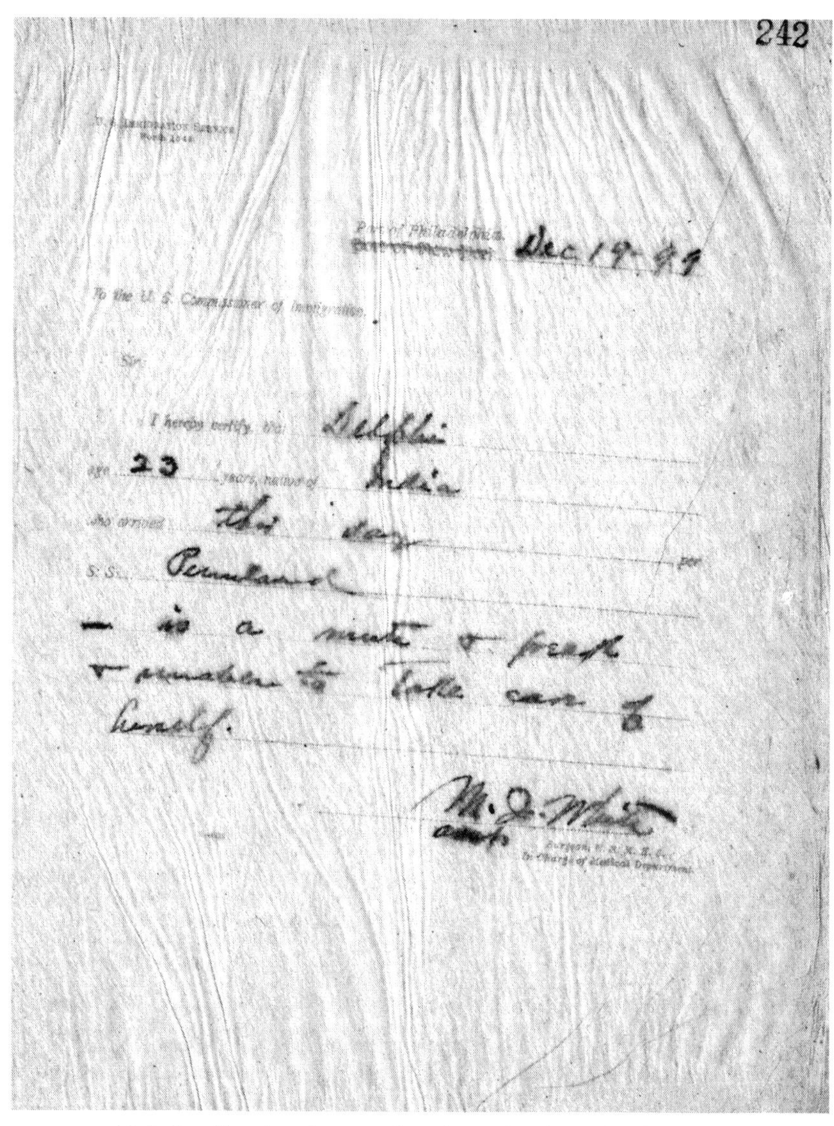

242

Port of Philadelphia, Dec 19 99

To the U. S. Commissioner of Immigration.

Sir:

I hereby certify, that Delphi

age 23 years, native of India

who arrived this day

S. S. Pennland per

is a mute & freak & unable to take care of himself.

M. J. White

FIGURE 1. Medical certificate issued to a member of a traveling show, identified here as Delphi, also known as Miss Delphi the Orange-Headed Girl. "I hereby certify that <u>Delphi</u>, age <u>23</u> years, native of <u>India</u>, who arrived <u>this day</u> per S.S. <u>Pennland</u>, ~~has~~ <u>is a mute and freak and unable to take care of himself</u>." The use of "freak" as a diagnosis shows the latitude Public Health Service physicians had to identify and describe excludable conditions. (National Archives, Records of the INS, Reports of Medical Inspectors in Philadelphia and New York, 1896–1903, RG 85, entry 2, vol. 1896–1900.)

used to justify inequality for not only disabled people but virtually every other group that has faced stigma or oppression. In the debates over women's rights in the nineteenth and early twentieth centuries, for example, opponents often cited women's purported incapacity for rational thought, their excessive emotionality, and their physical flaws and weaknesses. These are essentially mental, emotional, and physical disabilities, although historians have rarely discussed their significance as such or gone beyond condemning their imputation as slander. Opponents of racial equality often defined African Americans' supposed inferiority as a collection of defects, including a propensity to feeblemindedness, mental and physical illness, impaired reason, even deafness, blindness, and other disabilities resulting from "constitutional deficiencies." The cultural meanings associated with class, as well as age in every stage of life, have revolved around historically changing notions of disability, defect, and degeneracy. Sexuality has been deeply interconnected with notions of mental health and illness, which was why the decision of the American Psychiatric Association in 1973 to remove homosexuality from the *Diagnostic and Statistical Manual of Mental Disorders* was a matter of such great moment. The power of disability to discredit is evidenced by the typical response, a vigorous denial of any impairment: "We are not disabled," the accused insist, and therefore lay claim to equality. Rarely has any inhabitant of an oppressed category chosen instead to argue that disability is irrelevant to the question of equality and no justification for discrimination. Disabled people themselves form one of the minority groups historically assigned inferior status and subjected to discrimination, but disability has functioned for all such groups as a sign of and justification for differential and unequal treatment.[9]

The effort to bar disabled immigrants in the late nineteenth and early twentieth centuries was part of a larger story, one expression of a culture that was increasingly intolerant and afraid of difference, and that strove in sundry ways to keep disabled people apart from public life. In the United States, as in many industrialized countries, tens of thousands of people with psychiatric or intellectual disabilities were segregated into institutions, often under horrific conditions. Many thousands more were prohibited from marrying or involuntarily sterilized under state eugenics laws. Popular books, articles, even movies advocated (and some physicians aggressively practiced) the euthanasia of disabled infants.[10] A campaign to end the use of sign language by deaf people led to its prohibition in most classrooms in schools for the deaf and to generations of deaf adults who were ashamed to sign in public.[11] At the same time that widespread discrimination prevented many disabled people from earning a wage, cities enacted "unsightly beggar" ordinances to specifically

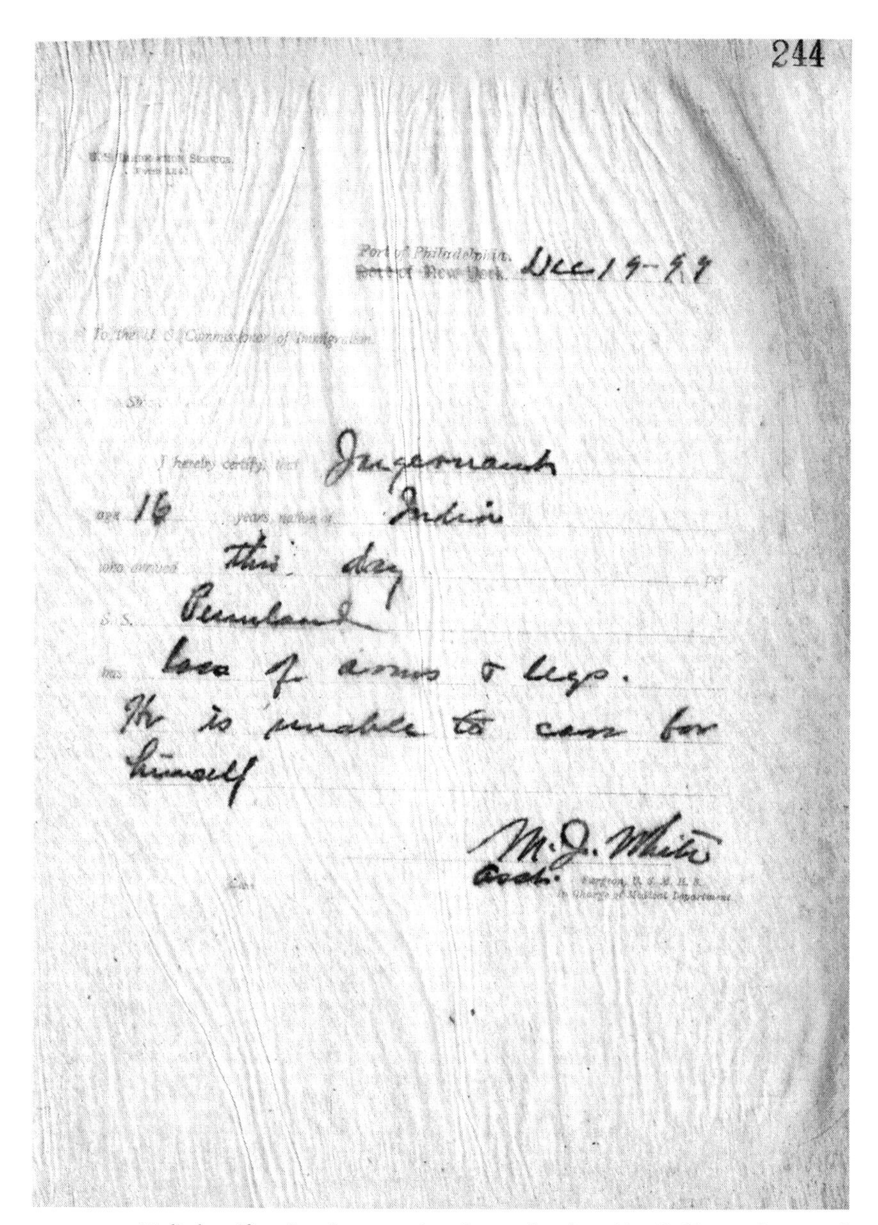
244

Port of Philadelphia. Dec 19-99

To the U. S. Commissioner of Immigration.

Sir:

I hereby certify, that Jugernaut

age 16 years, native of India

who arrived this day per

S. S. Pennland

has loss of arms & legs.
He is unable to care for
himself

M. J. White
Asst. Surgeon, U. S. M. H. S.
In Charge of Medical Department

FIGURE 2. Medical certificate issued to a member of a traveling show, identified here as "Jugernaut" [*sic*], also known as Juggernaut the Armless and Legless Mite. "I hereby certify that Jugernaut, age 16 years, who arrived this day per S.S. Pennland, has loss of arms and legs. He is unable to care for himself." (National Archives, Records of the INS, Reports of Medical Inspectors in Philadelphia and New York, 1896–1903, RG 85, entry 2, vol. 1896–1900.)

prohibit the "diseased, maimed, mutilated, or in any way deformed" from begging on city streets.[12] In short, as the historian Paul Longmore once observed, during these years "prejudice and discrimination against disabled people seem to have been intensifying sharply."[13]

Immigration restriction was the least ambiguous expression of this growing aversion to disability. Other measures primarily targeted specific types of disabilities and were defended, credibly or not, as being in the interests of those affected. Eugenic institutionalization, sterilization, marriage laws, even euthanasia were portrayed as benefiting not only the larger society but the affected individuals and their families. Restrictive immigration laws, on the other hand, eventually encompassed virtually all disabilities and were hardly defensible as beneficial to those who were refused entry. Thus, the course of immigration policy during the late nineteenth and early twentieth centuries offers an especially legible guide to that era's cultural assumptions about disability.

Chapter 1 lays the groundwork by showing that the primary concern of immigration restrictionists during these years was the exclusion of disabled persons, and by exploring the intersections between race and disability in immigration law. The concept of "selection," adapted from animal breeding and evolutionary science, was central to the eugenics project overall as well as specifically to immigration restriction. Eugenic selection of worthy citizens occurred along two main tracks. The one most vividly associated in the public imagination with the eugenics movement was the curtailment of reproduction by undesirable citizens through institutionalization, sterilization, marriage laws, and public education campaigns. Of equal importance, however, was the other main track of eugenic selection: the restriction of immigration by means of screening immigrants for defects.

Ideas about atypical bodies and minds clearly underwent dramatic changes in this era, but why? One crucial factor was a transformation in the meaning of time, the subject of chapter 2. Scientific discoveries in geology and evolutionary biology were radically altering concepts of historical time, while the growth of a market economy and industrial production were accelerating already changing perceptions of everyday time. Together, these shifts led to a redefining of disabled people as socially and economically "inefficient," drags on evolutionary and economic progress who were unable to keep up or successfully compete in the "race for life." New ideas and a new vocabulary for disability came into common use: "handicapped," "retarded," "abnormal," "degenerate," and others that were explicitly or implicitly rooted in new ways of thinking about time. Evolutionary theory was interpreted in ways that emphasized both the necessity and the practicability of seizing control of humanity's biological future. On the one hand, many expressed unbridled optimism that it was within the power of the present generation to commence building a physically, mentally, and morally superior human race. On the other were dark fears of a rising tide of hereditary defect, degeneration, and a dystopian future. The former showed the possibilities; the latter, why immediate action was imperative.

Economic anxieties arising from a largely unregulated market economy and rapid industrialization also affected how Americans understood the relative values of independence and dependence, which I explore in chapter 3. In a context of economic insecurity, the ideal of independence became ever more powerful, and disabled persons were increasingly described as dependent and burdensome. What had been earlier in the century primarily a family and community issue in the new economy became a social problem to be addressed at the level of the state and the nation. Would-be immigrants with disabilities confronted a widely shared presumption of dependency, one that informed the crafting of policy by lawmakers and its enforcement by immigration officials.

Chapter 4 explores the growing importance of appearances and first impressions in an increasingly urbanized and anonymous society and the mounting intolerance for difference. The everyday necessity for city dwellers to make quick appraisals of strangers, and for employers to select workers from among a surfeit of applicants, made them increasingly alert and sensitive to unusual bodies. Restrictionists worried greatly about the appearance of recent immigrants, describing them as malformed, undersized, oddly shaped, and ugly. Many immigrants who had no functional impairment were excluded as "likely to become a public charge," based solely on impaired appearances, by officials who thought them therefore unemployable as well as undesirable hereditary specimens. Immigration inspectors necessarily relied on first impressions, general appearance, and visible abnormalities as they scanned faces and bodies streaming past them for telltale signs, making what they termed "snapshot diagnoses." Visible defects stigmatized, but what inspired even greater fear was what could not be seen and might pass undetected. Germ theory and the spread of infectious diseases had brought to public consciousness the microscopic menace of bacteria and viruses; worse yet were the insidious dangers identified by the fledgling science of heredity. Defective "germ plasm," as it was called, threatened not merely illness in the present, but an ever-widening stream of polluted heredity flowing into the future.

A note on terminology and scope: It should be said at the outset that this is not a comprehensive history of immigration policy, even for the relatively brief span of time the book covers. My focus is on European immigration, most of it to Ellis Island and other ports in the northeast. Many important aspects of the political debate over restriction, immigration legislation, and the machinery of enforcement, are passed over briefly or entirely absent. I do not touch on the desire of business interests to maintain a flow of cheap labor, for example, or labor unions' corresponding opposition to it. Instead, what I have tried to do is to begin building a case that disability played an important role, even a central one, in the formation of early immigration law and policy in the United States, and that historians have heretofore overlooked its significance. I also propose some ideas to explain why disability came to seem a matter of such grave import to Americans at the time. Of course, I also hope that my underlying arguments will extend beyond the realm of immigration law to other fields of history, and to other disciplines than history.

I use the term "disability" in the introduction and conclusion, but in the main body of the book I keep more often to the terminology of the time. I discuss these terms and their contemporary significance as they arise. "Defect" was a key word, and in most usages corresponded closely to what we now mean by disability, although often applied more broadly and with connotations specific to the time. Although "hereditary defect" was sometimes specified, in many cases the term "defect" alone carried an implication of heritability. A far greater range of characteristics, strengths, weaknesses, and tendencies were considered heritable than are today, including, for many scientists and laypersons, characteristics acquired over the course of a life. Of the three broad categories of physical, mental, and moral defect, the first two are, for the most part, a comfortable fit with the way most people think about disability today. The third may seem less so. The concept of moral defect was part of an attempt to move beyond the language of sin and to explain deviant, criminal, and other socially unacceptable behaviors in biological terms. As the three general types of defect were thought to be deeply interconnected, moral defect was a frequent companion to mental defect and found outward expression not only in behavior, but in physical deficiencies as well. Moreover, one type of defect could mutate across generations into another. Thus, taking into account differences of historical context, the notion of moral defects fits comfortably under the rubric of disability. To anyone conversant with current psychiatric classifications of, for example, disruptive behavior disorders, personality disorders, alcohol behavior disorders, and impulse control disorders, understanding "moral defect" in terms of disability will not seem strange at all.

Like disability, race is a difficult concept to define with any precision, and even more so when trying to pin down how people talked about it in the past. The sociologists Michael Omi and Howard Winant, after struggling to find some workable definition for this "unstable and 'decentered' complex of social meanings constantly being transformed by political struggle," concluded that "race is a concept which signifies and symbolizes social conflicts and interests by referring to different types of human bodies."[14] This definition has the virtue of being broadly applicable but achieves that at the cost of being fairly nebulous. It might serve equally well as a definition for disability, a correspondence that suggests the interconnected meanings and uses of race and disability. Moreover, through the early twentieth century the concept of race in the United States often encompassed not only physical but also cultural difference—that is, it covered what became "ethnicity" after the 1920s.

Suffice it to say here that race in the late nineteenth and early twentieth centuries was understood differently and used more expansively than it is today; it was probably more potently charged, yet at the same time more flexible and ambiguous. In some contexts, such as in discussions of European immigration, "race" was used more or less interchangeably with "nationality." People commonly spoke of the Italian race, the Irish race, the British race, and so on. In other contexts, they spoke of races that shared territorial, linguistic, or migratory histories, such as the Teutonic, Anglo-Saxon, Alpine, Nordic, Mediterranean, Hebrew, or Slavic race. There was also an "American race," described as an admixture of European races, chiefly Anglo-Saxon and Teutonic, and shaped over generations by the particular conditions of life in America but facing the danger of "mongrelization" by recent arrivals from less favored races. When speaking of Europeans, race was usually understood as comprising both heredity ("blood") and culture, which were not thought to be distinct in the way they later would be, and was described by some experts as fixed and unchanging except over very long periods of time, and by others as susceptible to fairly rapid change under the influence of the environment. The European races were often ranked on a hierarchy that depended to a great extent on the supposed prevalence of hereditary defects within each. In other contexts, however, race could refer to the five great races of mankind, black, yellow, red, brown, and white. The same term was used, but carried profoundly different connotations. For these, blood was usually accorded greater weight than culture, differences were seen as more profound and resistant to change, and "race-mixing" was considered more calamitous and repugnant. My focus is on the attempts to regulate European immigration, and therefore mainly on the concept of race as applied to Europeans.[15]

The Immigration Act of 1882 was preceded by several months by the Chinese Exclusion Act (subsequently expanded to cover other Asian countries in 1917 and 1922). This and subsequent acts coincided with the end of Reconstruction and the reestablishment of a regime of white supremacy in the South. Making distinctions between white and nonwhite was fundamental to the evolving definitions of citizenship in the United States during these years. In the 1920s immigration quotas were put in place for the purpose of reducing immigration by the "undesirable races" of southern and eastern Europe. The quota laws not only drew a dividing line between the superior and the inferior races of Europe, but, as Mae Ngai has argued, "also divided Europe from the non-European world," for it "proceeded from the conviction that the American nation was, and should remain, a white nation."[16] In light of this context, reading forward from Chinese exclusion and backward from the quota acts, historians of immigration restriction have naturally looked to race as an analytical category for the intervening years. Nevertheless, however ambiguous the status and perceived "whiteness" of the disfavored races of Europe in the American imagination, as I will argue in the first chapter, disability was the primary preoccupation of those who wanted to reduce immigration from Europe. While it is certain that immigration restriction rested in part on a fear of "strangers in the land," in John Higham's phrase, it was also fueled by a deeper, more potent fear of defectives in the land.[17]

I want to thank friends and colleagues who have read and commented on previous drafts of the book, individual chapters, or papers from which chapters grew: Jenifer Barclay, Laurie Block, Ken Cmiel, Amy Fairchild, David Gerber, Nancy Hirschmann, Angela Keiser, Linda Kerber, Cathy Kudlick, Alan Kraut, Lawrence Levine, Beth Linker, Paul Longmore, Lucy Salyer, Kirilka Stavreva, James Trent, and Lauri Umansky. Thanks also to the archivists at the National Archives in Washington, DC, Philadelphia, and San Bruno, the Library of Congress, the Smithsonian Institution, and the American Philosophical Society, without whose knowledge and assistance this book could not have been written.

Portions of this book have appeared in different forms in the journals *Health and History*, *Sign Language Studies*, *Journal of American Ethnic History*, and in Paul K. Longmore and Lauri Umansky (eds.), *The New Disability History: American Perspectives* (New York University Press, 2001), and Nancy J. Hirschmann and Beth Linker (eds.), *Civil Disabilities: Citizenship, Membership, and Belonging* (University of Pennsylvania Press, 2015). I appreciate their permission to reprint this material.

1

Defective

Selection is a fraught word for people with disabilities. Such terms as "prenatal selection," "selective reproduction," and "genetic selection" raise the specter of disability *de*selection based on normative assumptions about what constitutes "a good life" or "a life worth living." Reproductive selection today is generally framed as an individual choice (although some ethicists maintain that it is an illusion that these kinds of decisions could exist apart from social norms and pressures).[1] Eugenicists in the late nineteenth and early twentieth centuries, however, frankly advocated reproductive selection as a collective decision, to be made at the state or national level, carried out by social pressure and persuasion when possible and by coercion when not. They advanced normative assumptions as scientific fact, confidently categorized human beings into types according to their economic and social value, and regarded the elimination of what they termed "defectives" as common sense.

The intentional improvement of animal stock, what today is usually referred to as "breeding," in the nineteenth century was termed "selection." It was because of the familiarity of the term that Charles Darwin settled on "natural selection" to describe the means by which evolutionary change occurred. Although he worried that the term was "in some respects a bad one, as it seems to imply conscious choice," he believed that its utility as an explanatory device outweighed that disadvantage because "it brings into connection the production of domestic races by man's power of selection, and the natural preservation of varieties and species in a state of nature."[2]

If Darwin began with selection as practiced by animal breeders to introduce the concept of natural selection, eugenicists went in the opposite direction, using natural selection to clarify the necessity of artificially selecting human stock. Natural selection, they explained, ruthlessly eliminated weaknesses and defects in nature, but under the conditions of modern civilization was defanged and declawed. Darwin himself had noted that we "we build asylums for the imbecile, the maimed, and the sick; we institute poor-laws," and as a result "the weak members of civilised societies propagate their kind." He attributed these actions, however, to "the noblest part of our nature," and avowed that "to neglect the weak and helpless" would gain merely "a contingent benefit, with a certain and great present evil." Therefore "we must bear without complaining" the weakest among us "surviving and propagating their kind."[3] To bear without complaining was not part of the eugenicist creed, however. Although actively preventing the survival of the weak was a line most (if not all) eugenicists were disinclined to cross, allowing them to propagate their kind was a different matter, and eugenicists believed that they had tools of selection at hand that were both ethical and practical.[4]

Eugenics was primarily a nationalistic project. Although eugenicists spoke about both the right of the individual to be "well-born" and the imperative that humanity progress, most of their attention focused on the middle ground of the nation. In the United States, as in many other countries, eugenicists carried on a passionate, decades-long debate over the most efficacious and least costly methods of selection for preserving and enhancing "a superior national race." When Henry Fairfield Osborn, Columbia University professor of zoology, wrote to the *New York Evening Journal* in 1911 to advocate more stringent inspection of immigrants, his opening line—"As a biologist as well as a patriot . . ."—perfectly captured the intersection of eugenics and nationalism.[5]

Eugenic selection of worthy citizens occurred along two main tracks. The one most vividly associated in the public imagination with the eugenics movement was the curtailment of reproduction by undesirable citizens through institutionalization, sterilization, marriage laws, and public education campaigns—fitter family and better baby contests, school and college courses, and a steady stream of articles, books, and sermons. Although federal courts became involved from time to time, this aspect of the eugenics movement was carried out mostly at the state and local level.[6] There was, however, another "field in which the federal government must cooperate," wrote Harry Laughlin, director of the Eugenics Record Office at Cold Spring Harbor, "if the human breeding stock in our population is to be purged of its defective parenthood."[7] That field, the other main track of eugenic selection, was the restriction of immigration by means of screening immigrants for defects.

Immigration restriction via selection was the most unambiguous expression of eugenic nationalism. Advocates of institutionalization and sterilization could include among their professed motives altruistic ones: to shelter vulnerable people, relieve parents of terrible burdens, prevent lives of presumed misery, and foster human progress. The popular writer Albert Wiggam was certain that Jesus, were he to return in the present day, would update the golden rule in light of modern eugenic science: "Do unto both the born and the unborn as you would have both the born and the unborn do unto you."[8] Eugenicists were humanitarians, according to a pamphlet from the American Eugenics Society, who did not have "less sympathy for the unfortunate" but rather wished "to alleviate their suffering, by seeing to it that everything possible is done to have fewer hereditary defectives."[9] The sociologist Charles Henderson believed that the state had an obligation to the "unfit" to actively "prevent their propagation of defects and thus the perpetuation of their misery in their offspring."[10] Harry Laughlin maintained that the "sum total of human freedom and human happiness will be greatly promoted [by] the elimination of degenerate and handicapped strains."[11] The physician Harry Haiselden claimed to advocate and practice the euthanasia of disabled infants because he "loves them," death being for them "the kindest mercy."[12] In contrast to these supposed ethical and moral obligations to segregate and sterilize, protecting the nation from defective immigrants, on the other hand, was never plausibly defensible as beneficial either to the individuals affected or to humanity at large. It promised only to keep defectives where they were: elsewhere.

Institutionalization and sterilization, moreover, affected a subset of disabled people, mainly those who fell into the categories of mentally or morally defective, and of those a minority. Deaf individuals were occasionally institutionalized, but because they were mistakenly thought to be mentally defective, not on account of their deafness.[13] People whose physical disabilities were rooted in brain disorders such as cerebral palsy and epilepsy were also sometimes institutionalized and sterilized as mental defectives. Proposals for state eugenic marriage laws occasionally included deaf, blind, and otherwise physically disabled people, but these did not become law in any state. Those that did become law, beginning with Connecticut in 1896, targeted persons considered to be mental and moral defectives.[14]

Immigration restrictions, by contrast, eventually encompassed virtually all varieties of disability. While mental and moral defects were considered the most serious and were mandatorily excluded under the law, a person with a disability of any kind was subject to heightened scrutiny and could be turned away on that basis. The idea that immigration policies should be based on eugenic principles was broadly popular. The *Portland Oregonian*, for example, headlined a story on the immigration service in 1913, "Government Stands as

Putting Our Immigrants Through the Sieve at Ellis Island

Government Stands as "Doctor of Eugenics" at Portals of Nation

Tests that Determine Who Shall and Who Shall Not Become Citizens of the United States Have Undergone Marked Changes—How Visual Tests Are Applied and the Mental and Physical Defectives Are Singled Out.

FIGURE 3. "Mentally and otherwise defective immigrants who might prove a menace to American manhood and womanhood are turned back." "Putting Our Immigrants through the Sieve at Ellis Island: Government Stands as 'Doctor of Eugenics' at Portals of Nation." *Portland Oregonian*, September 28, 1913. © Copyright, Oregonian Publishing Co.

'Doctor of Eugenics' at Portals of Nation," and touted the "marvelous system of practical eugenics" being put into action by immigrant officials who "have made practical eugenics a daily study." Americans "concerned about the maintenance and improvement of the mental and physical well-being of our race," the American Eugenics Society declared in 1914, were "turning more and more to the regulation of immigration as one of the most obvious means of accomplishing this purpose."[15]

The central task of immigrant medical inspection was to uncover defects, which potentially included any unwanted deviation from what was considered normal—mentally, morally, or physically. A defect might be a straightforward and visible impairment or an ill-defined degeneracy that manifested itself indirectly in multifarious ways—crime, illicit sexuality, or poverty, for instance. Defects of the body, mind, and moral sense were understood as deeply interconnected, which was what made them so worrisome. Eugene Talbot, a prominent Chicago surgeon and professor of medicine, for example, wrote in 1898 that most "crime is hereditary, a tendency which is, in most cases, associated with bodily defect, such as spinal deformities, stammering or other imperfect speech, club-foot, cleft-palate, hare-lip, deformed jaws and teeth, deaf-mutism, congenital blindness, paralysis, epilepsy, and scrofula."[16] George Lydston, a professor of medicine and criminal anthropology, went further and argued in 1904 that "defective physique" was not merely associated with criminality but a crucial factor "in the causation of crime," and reminded his readers of the "old adage, *mens sana in corpore sano*."[17] Francis Galton, originator of the term "eugenics," in 1905 advocated issuing eugenic certificates for "goodness of constitution, of physique, and of mental capacity," emphasizing that these were by no means "independent variables."[18] The nation's most famous physical educator, Dudley Sargent, in 1909 made the claim that "criminals, dullards, and the mentally defective" were all known to possess "very poor physiques."[19] And the director of an institution for the feebleminded warned in 1909 that "every imbecile, especially the high-grade (or brighter) imbecile, is a potential criminal, needing only the proper environment and opportunity for the development and expression of his criminal tendencies."[20] Immigration officials shared these assumptions. An Ellis Island physician in 1906 warned that "there is to be expected in the case of poor physique, as an accompaniment of signs of physical degeneracy, some abnormality in the individual's mental and moral make-up."[21]

Given the inchoate understanding of heredity, which even as late as the 1930s often included neo-Lamarckian ideas about the inheritance of acquired characteristics, defects were assumed to be not only heritable but also mutable, manifesting themselves in varied forms and having disastrous effects

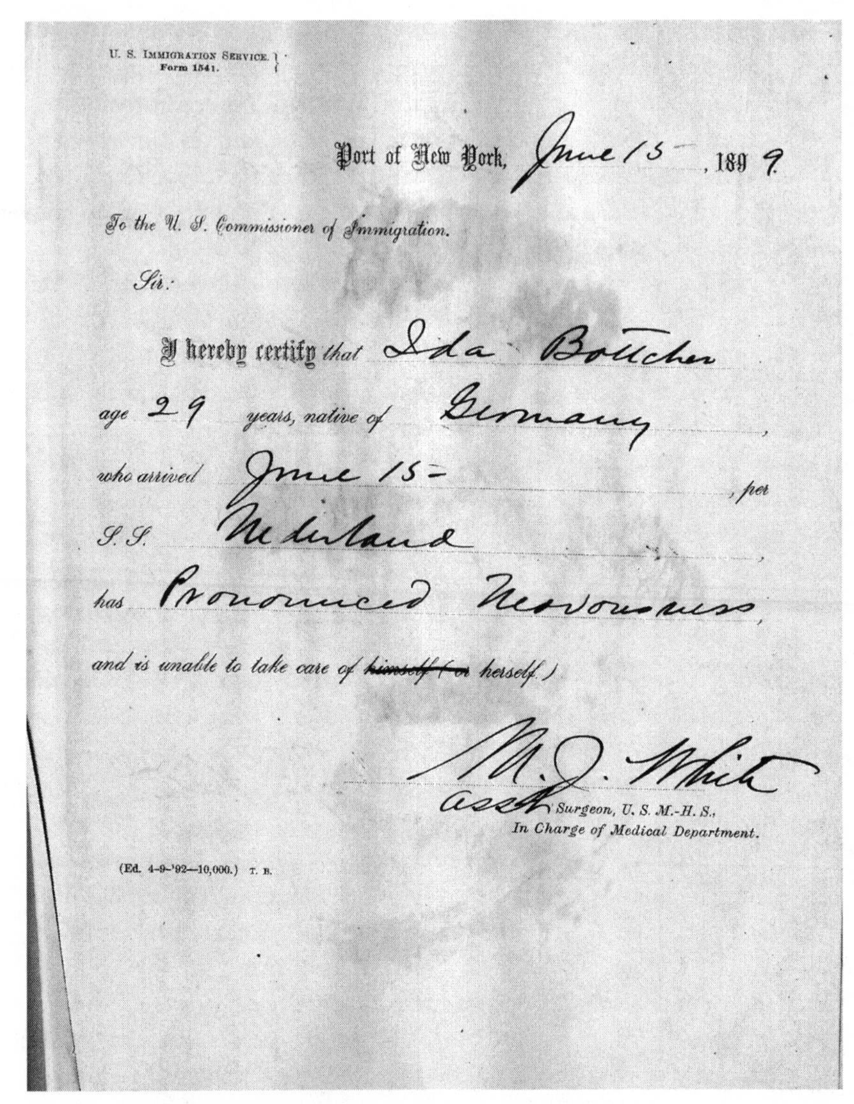

U. S. Immigration Service.
Form 1541.

Port of New York, June 15, 1899

To the U. S. Commissioner of Immigration.

Sir:

I hereby certify that Ida Bottcher

age 29 years, native of Germany,

who arrived June 15, per

S.S. Nederland,

has Pronounced Nervousness,

and is unable to take care of ~~himself (or~~ herself.)

M. J. White

asst Surgeon, U. S. M.-H. S.,
In Charge of Medical Department.

(Ed. 4-9-'92—10,000.) T. B.

FIGURE 4. "I hereby certify that Ida Bottcher, age 29 years, native of Germany, arrived this day per S.S. Nederland, has Pronounced Nervousness, and is unable to take care of ~~himself (or~~ herself)." Nervousness, also known as neurasthenia, was a diagnosis suggesting potential degeneracy. (National Archives, Records of the Immigration and Naturalization Service, Reports of Medical Inspectors in Philadelphia and New York, 1896–1903, RG 85, entry 2, vol. 1896–1900.)

on succeeding generations. The term for this phenomenon was "degeneracy," the tendency of defects to persist across generations, to mutate and metastasize, such that a mild defect might within several generations become a thoroughly corrupted nature. What was termed "moral imbecility," an "absence of the moral sense often as complete as is the absence of sight in the blind," was routinely associated with other defects: "Its influence in heredity is far-reaching, liable to reappear in its own or in another form of defect." Imbecility in all its forms had a "permeating, penetrating, disintegrating power" and was "at once the most insidious and the most aggressive of degenerative forces; attacking alike the physical, mental and moral nature, enfeebling the judgment and will, while exaggerating the sexual impulses."[22] An 1891 *Boston Medical and Surgical Journal* editorial could state without fear of being contradicted that "physical degeneracy is now known to go hand-in-hand with mental and moral degeneracy."[23] This was the critical threat to the nation that immigration restriction was meant to combat. The question was not merely whether the present generation of immigrants would make less desirable citizens, but whether they would degrade the quality of future generations of Americans.

Selection and Restriction

Historians have given relatively little attention to the significance of "defect" and "selection" in the history of immigration policy. The early years of immigration law are often divided into two phases: a selective phase starting with the Immigration Act of 1882 (with the Chinese Exclusion Act of that same year often cited as an exception to the general trend),[24] and a restrictive phase beginning either in 1917 with the passage of a literacy test or in 1921 with the first national quotas. The first phase is described as intended to screen out undesirable individuals, mainly those with physical, mental, and moral defects; the second, to target racial or ethnic groups and reduce the level of immigration overall. Selective laws are often minimized or depicted as reasonable efforts to protect the nation from harm, while the term "selection" is treated as transparent and straightforward, its charged significance in the rhetoric of eugenics overlooked. Restrictive laws, by contrast, are treated as more momentous, condemned as motivated by racism, nativism, and eugenics and subjected to far more critical inquiry.

Philip Taylor, for example, explained that despite agitation against unrestricted immigration, "very little was done" until national quotas were instituted in the 1920s; until that time, the laws "could reasonably be defended by arguments about anti-social behavior of a rather obvious kind, or physical and mental defects disqualifying the immigrant from any effective share in American life."[25] Roger Daniels wrote that "by 1917 the immigration policy of the United States had been restricted in seven major ways," affecting Asians, criminals, violators of moral standards, persons with diseases, paupers, radicals, and illiterates.[26] Disability was absent from a similar list by Bill Ong

Hing, who wrote that over the course of American history, "exclusionist rationales have been codified reflecting negative views toward particular races or nationalities, political views (e.g., Communists or anarchists), religions (e.g., Catholics, Jews, Muslims), or social groups (e.g., illiterates, homosexuals)."[27]

So much greater significance has been attributed to the national quota laws of the 1920s that the preceding four decades of restrictive laws at times virtually disappear from view. Categorical statements are not difficult to find: "In May 1921 the era of open immigration to the United States came to an abrupt end";[28] it was "1924 when free and open immigration ended";[29] "immigration policy was famously open until the 1920s, when eugenics arrived with a vengeance";[30] and "1924 marked both the end of one era, that of open immigration from Europe, and the beginning of a new one, the era of immigration restriction."[31] A chapter in a document collection on immigration is titled "Limited Naturalization, Unlimited Immigration—1880 to 1920."[32] One study suggested that "because race was so important to the *original opponents* to immigration, policymakers used racially motivated national quotas to restrict immigration," which makes sense only if the preceding laws are said to have been, as in this work, "symbolic."[33]

This period of ostensibly free, open, and unlimited immigration began in 1882 (following passage of the Chinese Exclusion Act, which, as historians of Asian immigration have often pointed out, was undoubtedly restrictive) by mandating the exclusion of any "lunatic, idiot, or any person unable to take care of himself or herself without becoming a public charge" (which applied to persons with physical disabilities). Determining the capacity for self-support was up to immigration officials, although their scope for judgment was narrowed in 1891 when "*likely* to become a public charge" became the criterion, and further still in 1907 when officials were directed to exclude anyone having a "mental or physical defect being of a nature which *may affect* the ability of such alien to earn a living." So-called lunatics and idiots were joined in 1903 by epileptics, as well as those who had been "insane within five years previous" or had experienced "two or more attacks of insanity at any time," and in 1907 by "imbeciles" and "feeble-minded persons."[34] In 1917 Congress thought it prudent to consider one previous "attack" of insanity sufficient cause, and to add people of "constitutional psychopathic inferiority," meaning the "various unstable individuals on the border line between sanity and insanity, such as moral imbeciles, pathological liars, many of the vagrants and cranks, and persons with abnormal sex instincts." Harry Laughlin approvingly called it a "scrap-basket" category that "implied poor stock in the family [and] degeneracy."[35] Inspection regulations that year directed the exclusion of those with "any mental abnormality whatever" as well as "aliens of a mentally inferior type."[36]

Public officials at the time did not see these laws as insignificant or merely symbolic. The commissioner general of immigration reported in 1907 that "the exclusion from this country of the morally, mentally, and physically deficient is the principal object to be accomplished by the immigration laws."[37] In 1910 the Ellis Island commissioner affirmed that "no part of the immigration law is more important" than those applying to mental defectives.[38] In 1911 another commissioner general reported that, for the American public, preventing "the entry into its midst of the physically, mentally, or morally unsound" was the issue "of prime importance now and always."[39] The report of 1919, by yet another commissioner general, noted that "one of the fundamental principles upon which our immigration legislation is based" is the exclusion of those who fall below a "normal standard, whether physically, mentally, or morally," and that the bureau "regards the effective enforcement of the law providing for exclusion and expulsion of such aliens as perhaps the most important of the trusts committed to it."[40]

Histories of immigration policy based on a distinction between selective and restrictive phases attribute to each its own purpose: the former to screen out (alternatively to sift, filter, strain, weed out, or winnow) undesirable persons and the latter to reduce numbers overall and more particularly those of disfavored races. John Higham's *Strangers in the Land*, for decades after its publication in 1955 the standard work on immigration policy, was structured along these lines. Higham wrote that during the 1880s Congress "put aside plans for reducing the absolute number of immigrants and concentrated instead on regulation and 'selection.'" When legislation mandating a literacy test for immigrants was passed by Congress but vetoed in 1907, he called it "the failure of restriction," concluding that no restrictive legislation "of any consequence became law for a decade after the essentially anti-restrictive measure of 1907."[41] This "anti-restrictive measure," in addition to prohibiting entry by idiots, imbeciles, feebleminded persons, epileptics, insane persons, and anyone with a history of insanity, created a special provision for persons with any kind of "mental or physical defect," required that ship surgeons inspect passengers before embarkation, and held ship captains responsible for affirming the absence of defective passengers.[42] The first law that Higham considered restrictive in intent, because it mandated a literacy test, came in 1917. The distinction for Higham was that earlier laws were intended to screen individuals, while the literacy test was implicitly meant to target groups by race (which the quota laws later did explicitly).[43]

Others have followed the same template, if not always the precise dates or terminology. Marion T. Bennett observed that the decades from 1880 to 1920 are "generally called the Selective Period of qualitative controls," followed by "the Restrictive Period of numerical or quantitative restriction."[44] Edward Prince Hutchinson agreed with Higham that the literacy test represented the "turning point in American immigration policy," when Congress made "a definite move from regulation to attempted restriction."[45] Roger Daniels declared the literacy test "the first significant general restriction ever passed."[46] Kenneth Ludmerer adopted a variation on this approach, arguing that until the 1920s "federal policy embodied an economic, not a biological, view of immigration." The terminology was somewhat different, but the argument was based on the same grounds: that the earlier laws prohibited "the entry of individuals (but not races) deemed undesirable—criminals, polygamists, anarchists, and the feebleminded." Leaving aside the incompleteness of this list, which omits many of the defects that were cause for exclusion, the premise of his explanation is that when defective individuals were the perceived threat, exclusionary policies were based on rational economic considerations; when the threat was degenerate races, however, policies were based on irrational prejudice rooted in eugenic thinking.[47]

The "public charge" provision was represented at the time as a straightforward question of ability to work and to avoid dependence on public assistance, and historians generally take this representation at face value. Immigration officials never sought evidence, however, that disabilities made people more likely to become public charges, and if so, which particular disabilities. It was at heart not an empirical claim but a presumption that disability meant an inability to live independently.[48] Women were similarly considered dependents, despite the legions of working women and families dependent on them, and were often rejected on that basis by immigration officials when they arrived unaccompanied by a male provider.[49] In many cases, the defect for which immigrants were excluded entailed no impairment of function. For example, Donabet Mousekian, a photographer and Armenian refugee from Turkey, was rejected in 1905 for what the physician termed "feminism." As Mousekian said, "It won't do any harm to my working; what harm can I do to the U.S. by my being deprived of male organs?"[50] Others had been self-supporting in their home countries; for example, Adrianus Boer, a skilled leather worker and saddle maker from the Netherlands, was rejected in 1905 for deafness in spite of having "always worked and supported the family without any outside help at all."[51] In some cases, an immigrant received a job offer before being deported; in 1913, the brother of Moische

Fischmann brought a written job offer from his employer to Ellis Island, but Fischmann was deaf and so was turned away.[52] In many instances, immigrants had family willing and able to support them if needed, such as Helena Bartnikowska, who was refused entry in 1908 because "this supposed woman" was "a hermaphrodite."[53]

More important, to the extent that some people with disabilities might indeed have encountered difficulties in finding work, a simple economic explanation for their exclusion locates the problem of unemployment among disabled people in the bodies of individuals rather than in the relationship between particular bodies and the constructed physical and social environments in which they live. Whether or not legislators and immigration officials at the time should be faulted for failing to recognize disability as a social issue, historians should put immigration restriction into context as one element in a larger system of discrimination that made it difficult for disabled people to live and move about independently. The reasons for unemployment among people with particular disabilities at particular moments in history are questions to be argued and demonstrated, not simply assumed. The design of industrial jobs is not a natural but a social artifact. As Martha Nussbaum observes, "The well-being of people with minority impairments has rarely been considered in the design of buildings, communications facilities, and public accommodations." This historical context is rarely considered; although built environments "are in no sense inevitable or natural," they are made to seem so by pervasive "fictions of normalcy."[54]

The possibilities for workplace accommodations may be better understood today, but they were hardly unknown at the time. Professionals in the growing rehabilitation movement condemned the "social arrangement that virtually condemns the cripple to mendicancy."[55] Henry Ford declared in 1914 that Americans "are too ready to assume without investigation that the full possession of faculties is a condition requisite to the best performance of all jobs."[56] He announced a nondiscrimination policy for disabled workers and promptly hired thousands. The Red Cross Institute for Crippled and Disabled Men in New York City reported great success in finding employment for disabled workers in 1918.[57] The efficiency experts Frank and Lillian Gilbreth in 1920 advocated workplace reforms to "adapt the work to the man" by "rearranging the surroundings, equipment and tools . . . , modifications of machinery, [or] changing the method by which the work is done." The chief obstacle, they argued, was the public's view "that the thought of a cripple reentering competitive industrial life is repellent, that these people should be provided for by pensions in their homes." Economic concerns were clearly part of the debate, but the argument that a selective policy during the early years of immigration law arose simply from practical economic considerations does not fit the evidence.[58]

The number of those excluded each year specifically for disability is impossible to calculate because most were placed in the catchall category "likely to become a public charge." This was always the largest category of exclusion, but the criteria applied were never straightforward or clear-cut. Possession of a marketable skill and cash both counted favorably, but lack of either was no bar. In 1900, a correspondent for the *Chicago Inter Ocean* sat in on a Board of Special Inquiry hearing at Ellis Island and was amused when a young lion trainer from Burma came before them: "The boy has no money. What can he do? Tame lions. The board looks helpless. It don't know of any lions in its immediate circle of acquaintances." The lion tamer was admitted, nonetheless. Poverty in itself was not pivotal, for most immigrants were poor. Because life histories were difficult or impossible to confirm, finding ways to predict future poverty and dependence was the key. For immigration inspectors responsible for examining hundreds or thousands of immigrants in a day, disability was often the only, and always the most visible, sign.[59]

Most of those engaged in the immigration debate at the time did not make a distinction between selective and restrictive laws. In some cases they used the terms interchangeably, but most often they described immigration law as intended to accomplish restriction by means of selection. The 1882 act was commonly described as "restrictive" or "mildly restrictive" by contemporaries.[60] Following passage of the acts of 1891 and 1893, Bureau of Immigration annual reports spoke of "restrictive features" of the law "framed to sift the incomers."[61] The report at the end of 1893 explained a brief surge in immigration earlier in the year as the result of immigrants hurrying "to arrive before the restrictive measures" of the new law took effect,[62] and the following year's report attributed the subsequent decline to "the efficient execution of the immigration laws." It furthermore predicted that with improvements to the "system of inspection," both the "volume of immigration will be restricted" and its quality "will continue to improve."[63] The report of 1897 trumpeted a decrease "in the annual average of nearly 200,000" since 1891, "for which credit, in large part, must be given to restrictive legislation," and it expressed confidence that "with the present laws energetically enforced" it was unlikely that "immigration will ever reach the volume of past years."[64]

In short, the laws were functioning as designed—simultaneously selective and restrictive—to keep out defective individuals, nudge the racial balance of immigration in a satisfactory direction, and reduce total immigration. The bureau's understanding of how the laws were working was not idiosyncratic:

in 1892 the president of the New York Chamber of Commerce lauded the "restrictive laws" that excluded "lunatics, idiots," and others before calling for a literacy test to further restrict the number of immigrants and "improve their quality." In 1893 the chairman of the Senate Committee on Immigration recommended "further measures of restriction," and a year later the commissioner of Ellis Island pronounced the "sifting" and "winnowing" under the recent law "a great success as a restrictive measure."[65] A writer in the *Atlantic Monthly* backed him up, affirming that the efficacy of the law in "limiting the number and determining the quality of immigrants" had been "conclusively proved," and in 1896 that great champion of restriction, Senator Henry Cabot Lodge, introduced a bill "intended to amend the existing law so as to *restrict still further* immigration to the United States." [66]

The Literacy Test

As it later turned out, screening for defects did not live up to the hopes of the restrictionists: immigration began to increase again after the turn of the century, and the market for new ideas to address the problem continued to be a lively one. Some advocated placing a numerical cap on annual immigration or charging immigrants a head tax sufficient to deter those of limited means. Others favored a physical examination, perhaps modeled on that used by the military. The proposal that was eventually made into law was the literacy test. Grover Cleveland vetoed a literacy test bill in 1897, saying that the nation had always welcomed immigrants "except those whose moral or physical condition or history threatened danger to our national welfare and safety." The only reasonable requirement was "physical and moral soundness and a willingness and ability to work." He conceded the "necessity of protecting our population against degeneration," but restrictions that targeted the "unfit" more precisely were the appropriate action. If a "particular element of our illiterate immigration is to be feared for other causes than illiteracy, these causes should be dealt with directly, instead of making illiteracy the pretext for exclusion, to the detriment of other illiterate immigrants against whom the real cause of complaint can not be alleged."[67] In 1916, when a congressional committee reported on a literacy bill, the minority report raised similar objections, agreeing that "the criminal, the insane and the mentally defective, the morally unfit, the pauper, and those who are incapable of earning a living" ought to be barred. A literacy test, however, was not a "test of fitness," and it would "inflict an unjust hardship" upon those whom "we can not class among the undesirable." The literacy test was too blunt an instrument.[68]

After twenty years of failed attempts, a literacy requirement was finally enacted in 1917, over a veto by Woodrow Wilson (who had also vetoed a previous version in 1915). His veto messages may be one source for the interpretation that this law represented a turning point from selection to restriction. Wilson charged that such a law would constitute a "radical change in the policy of the nation" from an assessment of "quality" and "personal fitness" to one of educational opportunity. The nation had always "generously kept our doors open to all who were not unfitted by reason of disease or incapacity," but the goal of literacy-test advocates was "restriction, not selection." Wilson spoke assuredly of a consensus that selection on the basis of fitness could "not be objected to on principle," for excluding defectives was broadly popular. Restriction, on the other hand, was a more loaded term that for some suggested anti-immigrant animus. Most members of Congress, however, understood the literacy test as just another selective screen.[69]

Another plausible reason to place the literacy test under the mantle of restriction might be its close association with the Immigration Restriction League, which had promoted the test ever since its founding in Boston in 1894. The goal of the organization was proclaimed in its name, its publications argued that existing law was inadequate, and some of those publications claimed a literacy test would reduce overall numbers by barring many immigrants from inferior races. Here the literacy test and restriction might seem to intersect. More often, however, league publications promoted the test as more practical and effective but nevertheless an extension of earlier methods of restriction. Its constitution defined the goal of the organization as "further judicious restriction" to build on what already existed.[70] Resolutions shortly after its founding continued to call for "further restriction and stricter regulation."[71] A 1914 pamphlet argued that "the reading test is the best selective test," because it would "exclude a considerable proportion of defectives" as well as "many not absolutely defective or delinquent, but likely to become so." Furthermore, the test would complement and improve the existing system of inspection, for "by diminishing the number of doubtful cases it gives more time to the inspectors . . . to examine the remaining cases." For these reasons, it would be "far more selective than any other test proposed." It made no claims for substantially reducing immigration; instead, the pamphlet claimed that "it is certain to improve the quality of the aliens; but it is not certain to very largely diminish their numbers."[72]

The two principal founders of the league, Prescott Hall and Robert DeCourcy Ward, both described their support for the literacy test as a question of improved efficacy rather than a change in direction. In 1904 Hall wrote that the laws as they stood were fine in theory but "practically" incapable of solving the "main problem of the proper selection of immigrants." In search of "some further method of selection," he outlined several options before concluding that the literacy test would be the most efficient and practicable.[73] Writing that same year, Ward also favored "further restriction" with a new method for "selecting" desirable immigrants. He was impressed by "what has been accomplished" by Congress so far but emphasized that there was an important distinction between "defining what classes of immigrants shall be excluded" and creating "the machinery whereby the exclusion is accomplished." The "machinery," that is, the "inspection of the incoming aliens on the dock," was inadequate to the task. The literacy test might be the most workable solution, "not because illiteracy *necessarily* means that an immigrant will prove a bad citizen, but because the measure will be practical."[74] In the eyes of its proponents, the proposal represented no radical shift. As the commissioner general maintained in 1903, the adoption of literacy testing would be "merely additional steps in the same direction already taken" and "involve no new departure from a policy which has been pursued for years."[75]

Race

Historians have framed the literacy test as a turning point in immigration policy in part because they have seen it as a proxy for race. John Higham, for example, wrote that for most of its advocates, "it was chiefly a means of discriminating against 'alien races,'" and that they preferred it over other proposals, such as inspection at foreign consulates before embarkation, because the latter "would not discriminate between nationalities."[76] According to Barbara Miller Solomon, the goal was "preventing any further inroads on Anglo-Saxon America by strangers."[77] That the test *was* intended as a proxy for other matters its advocates frequently and openly conceded at the time. What it stood in for, however, was not so much nationality or race as the eugenic fitness of individuals.[78] When they invoked race, it was usually in the context of relative tendency toward defect, for it was commonly said that "illiterate races are generally inferior in physique" as well as prone to insanity and fits of violence.[79] For example, Hall insisted that when discussing the effects of immigrants on society "it is convenient to consider them by nationalities," but in practice "each individual must be judged on his own merits without race proscription or prejudice."[80]

Evolutionary thinking of the time posited two levels of competition: one in which individuals within a species competed with each other for survival, and another in which groups—communities, classes, nations, and races—competed. Eugenicists generally believed that within each group the fittest individuals would rise to the top, and that the upper stratum of any race was likely to be superior to the lower strata of any other. Thus, if the best of other nations came to the United States, that would enhance its population and its standing in the competition of nations. Ward argued that with appropriate policies, "we can pick out the best specimens of each race to be our own fellow-citizens and to be the parents of our future citizens," and he took pains to stress that "it should be most emphatically stated that no one should object to the coming of the better classes of Italians, Austrians, and Russians, even in fairly large numbers."[81]

Ward and other restrictionists did talk a great deal about race, of course. Ward argued in 1904, for example, that "we have spent too much time studying the economic sides of the question," and that there was "a racial side which is even more important than all the economic aspects put together." This would seem to suggest that he would go on to advocate restriction on the basis of race; instead, however, he urged more effective methods of individual selection. When he wrote that the "question before us is a race question," Ward meant that in the urban and industrialized United States, inferior types outbred superior types, that the "highest stratum" would eventually be "eliminated in the course of natural selection," and that this would "profoundly affect the character of the future American race." It was not a problem of foreign races per se so much as it was the decline of the American race owing to an influx of inferior individuals. He believed that the debate had focused too much on short-term economic risks posed by defective immigrants—those likely to become public charges—and not enough on the long-term genetic fitness of the nation.[82] In 1912 he argued that "the New England country towns are full of hopelessly degenerate native [born] Americans who are inferior mentally, morally and physically to the sturdy peasants of Europe." These were white Anglo-Saxons whose ancestors had immigrated generations ago. It was not race that underlay their inferiority. Instead, it was "probably chiefly due to the drawing off of the stronger and more capable men and women to the cities," a problem of adverse selection that led to the "degeneracy of our country native stock." Ward's point was that this made the rigorous selection of immigrants even more important, for it "will not help to reduce the number of our native degenerates if we admit alien degenerates."[83]

Ward may well have had in mind a study published that same year by Henry Herbert Goddard, a eugenicist, psychologist, and early advocate of the use of the Binet-Simon intelligence test to construct a diagnostic and classification system for intellectual disability. His book *The Kallikak Family: A Study in the Heredity of Feeble-Mindedness* purported to trace the defective lineage of a woman who was confined at the New Jersey Home for the Education and Care of Feebleminded Children, where Goddard was research director. Goddard's study was frequently cited in arguments for greater immigration restriction, yet, as Leila Zenderland has pointed out, it had almost nothing to say about race. Indeed, the Kallikaks were white Anglo-Saxon Protestants whose ancestors dated back at least to the American Revolution.[84]

The promise and peril of the American race, the "American of tomorrow," the race that was always in the making, figured far more conspicuously in restrictionist rhetoric than attacks on inferior European races. Although they thought that certain of the European races contributed a disproportionate share of defective blood to American veins, the primary problem was individual defect. In 1910 Commissioner General Daniel Keefe emphasized that in order to "preserve and improve the American race," it was important to remember that the "strength of a nation is the combined strength of its individual members." The American race may be an amalgamation of individuals from all over the world, but "if we continue to inject into the veins of our nation the blood of ill-formed, undersized persons, as are so many of the immigrants now coming," was it reasonable to expect that "the American of to-morrow will be the sturdy man that he is to-day?"[85] As the popular writer Montaville Flowers observed in 1917, "Citizenship in our land is not primarily a question of race, but it is a question of individual fitness." At the same time, he conceded that individual fitness "may be determined by the national and racial inheritances of their fatherlands."[86] Those inheritances mattered, but only secondarily. Their significance lay in the fact that they might determine the primary matter of fitness.

When historians have considered charges of "racial unfitness" leveled against immigrants, they have attended almost entirely to the "racial" part of the quotation and hardly at all to "unfitness." But the noun was more important than its modifier. Race was often used as a marker for unfitness, but did not on its own signify unfitness, and it was only one such marker among many others. Not all members of any inferior race were unfit, nor were all members of any superior race fit. It was instead a question of tendencies, percentages, and likelihood. Certain races invited greater skepticism and scrutiny, but none was uniformly one thing or another. The American race was made, after all, not just by direct inheritance from Europe, but also by the selective processes entailed by migration and pioneer life on the frontier. It was not uniformly fit and energetic, but this was believed to be its tendency overall. Race was a generalized abstraction. It was not this imagined category that finally held promise or danger, but rather the actual fit or defective individuals who composed it.

Disavowals of racist intent cannot simply be taken at face value, of course, but when coupled with statements of prejudice on a different basis that fit the evidence and explain actions just as plausibly or more so, they should be taken seriously. Eugenicist arguments more or less consistently emphasized individual defect over race. Charles Davenport, founder of the Eugenics Records Office at Cold Spring Harbor and a leading eugenicist, for example, insisted that the "old idea that there is in society any *class* that is superior to any other class should be abandoned," for it was "the characteristics of the germ plasm and not individuals as a whole that are favorable or prejudicial to human society." In regard to reproduction, the "way to improve the race is first to get facts as to the inheritance of different characteristics and then by acquainting people with the facts lead them to make for themselves suitable matings."[87] He reiterated this focus on individual heredity in regard to immigration in 1911, writing that "no race *per se*, whether Slovak, Ruthenian, Turk, or Chinese, is dangerous and none undesirable; but only those *individuals* whose somatic traits or germinal determiners" were undesirable. He favored not restriction by race, but rather investigation into the pedigrees of prospective immigrants: "if increasing attention is paid to the selective elimination at our ports of entry of the actually undesirable (those with a germ plasm that has imbecile, epileptic, insane, criminalistic, alcoholic, and sexually immoral tendencies)," then in fact "we may see our population not harmed but improved by this mixture of a more mercurial people."[88] In 1913 he declared his opposition to the literacy test, deeming it "unwise to insist on the exclusion of the illiterate, the indigent, the dark-skinned or the oblique-eyed per se." Instead, he favored "the most direct way" of excluding the unfit: "if the family strain is 'bad' he is to be excluded."[89] In 1920, at a meeting of the Eugenics Research Association Committee on Immigration, Davenport proposed a resolution that began "Whereas the protection of the germ-plasm of the nation is of prime importance for the future of the United States," went on to declare that "no mere quantitative method of selecting immigrants should be permitted to replace selection by quality," and concluded by calling on Congress to require that prospective immigrants furnish family histories before being approved for travel to the United States. This would keep out those who have "not only undesirable personal qualities but also family tendencies of physical or mental defect."[90] In a private communication to his fellow eugenicist Madison Grant, he wrote, "I am inclined to think that the thing to do is to make better selection of immigrants, to admit them in fairly large numbers so long as we can sift out the defective strains."[91] This was the primary objective for Davenport, as for most eugenicists and restrictionists, including proponents and opponents alike of the literacy test—to weed out defective individuals.

Many individual restrictionists were deeply interested in questions of race, but the correlation does not explain restrictionism. In practice, eugenic and racial ideas pervasive in American society could not help but mutually inflect each other in the immigration debate, just as both did with sexism and classism, but they were nonetheless distinct phenomena. For example, a *New York Times* editorial questioned why eugenicists were supporting the 1924 quota bill then before Congress: "For the eugenicist there are, at least in theory, no races. There are only the well-born and the ill-born. In every race only a small percentage carries the germ plasm that creates civilization and lifts the human race upward. In every race the great mass is, eugenically speaking, so much deadweight or worse."[92] Roswell Johnson responded to the editorial on behalf of the Eugenics Committee of the United States to point out that the "committee favors the Johnson [quota] bill only as a step forward," not as an ideal solution: "no bill that makes so much of race difference and so little of individual differences goes as far as eugenicists would wish it to go." Their support for national origin quotas was merely "a matter of practical legislation," he insisted, but if he and other "eugenicists could be assured that an effective individual selection . . . would be put in force and kept in force, we would gladly give up all national discriminations." The problem was simply that, as yet, "Congress and the American people lack that clear-cut conception of the situation." Therefore, for the time being, "let us keep the quotas and the literacy test until we have the newer and sounder policies" for stringent individual selection.[93]

Poor Physique

A criticism often directed against the literacy test was that it would bar many desirable along with undesirable immigrants. Advocates readily conceded the point. Illiteracy was not the main issue. Instead, the test was explicitly intended to be a proxy for screening out defective individuals. That this was so is evidenced by the other proposal most advocated by restrictionists and by the Immigration Restriction League itself: a strict physical standard for admission. Proposals for a physical test had come up periodically. The commissioner general in 1898, for example, recommended physical standards similar to those used by the military, arguing that if it was important to reject the "unsound in mind and body" when only a three-year enlistment was at stake, how much more so it was in the case of those "permanently enlisting in the army of producers in the United States and, what is far more important, of becoming the parents of future soldiers and citizens of this country." A physical standard, he thought, would provide "intelligent, discriminating restriction."[94]

The idea gained traction when William Williams, commissioner of the Port of New York at Ellis Island, took up the cause in 1903. Over the next several years Williams used his lectures to university and civic groups, articles in the *New York Times* and elsewhere, and his annual reports as commissioner to advocate a physical exam that would "reach a large body of immigrants" who are "of low vitality, of poor physique . . . and unfitted mentally or morally for good citizenship." The problem was that, under existing law, excluding defectives still required a finding that they were likely to become public charges, but "it is obviously impossible to exclude on this ground all persons whose physical condition is poor." His concern was not primarily the economic question of self-support. He granted that the laws already screened out the "mere scum or refuse, persons whom no country could possibly want," but further measures were needed to eliminate those who, while not clearly "riff-raff," were still far from desirable. His answer was to make immigrants of "poor physique" excludable on that basis alone, without the necessity of claiming that capacity for self-support was impaired. This would significantly reduce immigration, he argued, especially among the "undesirable and unintelligent people from Southern and Eastern Europe,"—or, as he described them, "the refuse of Europe." Like most other restrictionists, he emphasized that not all immigrants from southern and eastern Europe were undesirable. Rather, the problem was the defective individuals who so often came from that part of the world.[95]

The next immigration official to take up the cause was Allan McLaughlin, a Public Health Service physician at Ellis Island who outlined, at a meeting of the American Public Health Association in October 1903, the three main dangers immigration posed for public health. Of the three, he felt certain that the "physique of the immigrant . . . is by far the most important factor." The second was infectious disease, to which persons of poor physique, he noted, were "ready prey." Finally, there were the overcrowded cities, where "aliens of a race having a low physical standard will invariably herd together," as opposed to the "sturdy races," who "tend to establish little homes of their own in the country or in the suburbs." Three dangers, but one at the root of them all. The "real danger to public health" was that "we admit the immigrant with poor physique, unless it is so marked as to make him undeniably a public charge." McLaughlin's "sturdy races" rhetoric and his tables showing "the relative physical strength of the various races" at first glance suggests race as the primary concern. But when he came to prescribing remedies, he rejected "unjust discrimination against any particular race" as "illogical, bigoted, and un-American." The real issue for him was that the "indiscriminate admission of a horde of diseased, defective, and destitute immigrants would be a crime against the body politic." The logical remedy, then, would be "one standard of physique," an absolute directive to make "the medical certificate of poor physique equivalent to deportation." McLaughlin saw little merit in a literacy test save for "one good result" that would come of it: that it "would lessen the total number of immigrants landed and thus permit of an even more rigid examination of the immigrants upon arrival."[96]

Over the next several years poor physique moved to the center of the immigration debate, aided by articles in the popular press and reports, memoranda, and letters that circulated among immigration officials and restrictionist organizations. Williams's annual report was excerpted in the *New York Times*. McLaughlin's paper was reprinted in *Popular Science Monthly* in 1904, in *Public Opinion* in 1905, and in Robert La Follette's *The Making of America* in 1907.[97] McLaughlin followed up with several more articles in *Popular Science Monthly* making essentially the same point in various ways.[98] In one he declared that "the first requisite of a desirable immigrant is good physique," not necessarily because poor physique made one incapable of work, but because "even if he were able, employers of labor would not hire him."[99] If examiners believed an immigrant was likely to encounter discrimination on the basis of appearance, that too should justify exclusion.

In January 1905 Ward wrote to Commissioner General Frank Sargent to urge that Williams's proposal be "embodied in a general Bill to be introduced into the next Congress."[100] Congressional action on the issue, however, was uncertain and in any case would take time. Meanwhile the bureau took steps to do what they could within existing law. In March Sargent asked the head physician at the Baltimore station, Joseph W. Schereschewsky, to compose a memorandum for distribution to all of the immigration stations clarifying the threat posed by immigrants of poor physique. Schereschewsky replied (with impressive syntax) that a "certificate of this nature implies that the alien concerned is afflicted with a body but illy [*sic*] adapted not only to the work necessary to earn his bread, but is also but poorly able to withstand the onslaught of disease. It means that he is undersized, poorly developed, with feeble heart action, arteries below the standard size; that he is physically degenerate, and as such, not only unlikely to become a desirable citizen, but also very likely to transmit his undesirable qualities to his offspring, should he unfortunately for the country in which he is domiciled, have any."[101]

Being "illy adapted" to work and likely to fall sick could both be considered as coming under the public-charge provision. The description of hereditary impact, however, referenced an ever-present eugenic fear that was explicit in public debates over immigration law but an unwritten subtext to the law. Schereschewsky reinforced the message by adding that "of all causes for rejection, outside of those for dangerous contagious or loathsome diseases, or for mental disease, that of 'Poor Physique' should receive the most weight." The chief danger posed by these immigrants was that they would become "progenitors . . . whose offspring will reproduce, often in an exaggerated degree, the physical degeneracy of their parents."[102]

On April 12 the *Boston Evening Transcript* reported that societies for the protection of Jewish immigrants were protesting that "the recent practice in excluding persons because of 'poor physique' works a discrimination against the Jews." Orthodox Jews, they pointed out, could not eat the food provided onboard but had to "subsist on bread of their own and herring" for up to two weeks, which "makes them look in poor physique."[103] When Ward read the story, he hurried that same day to write to the Assistant Secretary of Commerce and Labor, Lawrence O. Murray, to tell him of his "satisfaction" in reading the news that the department was finally beginning to exclude aliens of poor physique, the "most important step which we can possibly take in the matter of a further selection of immigration." Immigrants of poor physique posed "the greatest danger to the health and strength of our future American stock," and he looked forward to Congress enacting an "amendment to our existing immigration laws which would make it *obligatory*" to exclude them.[104]

On April 17 Sargent circulated Schereschewsky's memorandum among the immigration stations "for the information and guidance of all officers," adding his own caution that "aliens so certified are particularly disqualified from admission to the United States, and that fact should be borne in mind by all officers who are called upon to determine the admissibility of arriving aliens." He concluded that all port commissioners "are instructed to disseminate this opinion of the Bureau to the fullest extent possible." Sargent then also responded personally to Ward, reiterating that the Bureau well understood "the necessity for rigid enforcement of the immigration laws," and enclosed a copy of the memorandum, hoping that it "will indicate the attitude of the Bureau with respect to aliens certified to be afflicted with poor physique."[105] Sargent's circular had a powerful effect. Although the diagnostic category was ill-defined, immigration officials gave it great weight, and it quickly became one of the most commonly issued medical certificates.[106]

At this unpropitious moment in history, Israel Bosak arrived at Ellis Island. Bosak was a tailor whose shop in Russia had been destroyed by a mob during an anti-Jewish pogrom. Arriving in January 1906, with $65 in his pocket (more than most immigrants had when they arrived), he was certified as having a poor physique and sent before the Board of Special Inquiry, a panel of three immigration officials that would determine his eligibility to land. At his hearing Bosak told the officials that he intended to establish himself in New York City and then to send for his wife and children. Two relatives came to the hearing to offer assurances that they would provide for Bosak until he had an income, help him to find a job, and assist his family when they arrived. Bosak explained that he also had "plenty of countrymen here who are just as good as relatives, to help me."

Two inspectors quickly voted to exclude on the basis of the certificate, but a new member of the board, Philip Cowen, disagreed. Cowen was a prominent writer on anti-Semitism, publisher of the New York weekly magazine *The American Hebrew*, and frequent advocate for immigrants. He had been appointed to the service by executive order of Theodore Roosevelt less than a year before. Later, in the summer of that year, Roosevelt would send him to Russia to investigate the conditions behind the large migration of Jews out of Russia. During his time as a member of the Board of Special Inquiry, he proved to be a maverick, often dissenting from the decisions of his fellow board members. After questioning Bosak about his tailor shop and the pogrom in which it was destroyed, he proceeded to make a speech (an unusual thing in itself at these hearings). He argued that facts usually considered irrelevant—that Bosak was "driven from his home by the mob" that had "despoiled him of the property" and "prevented him from earning a livelihood"—should be taken into account. He further maintained that Bosak "seems to me to be valuable material for immigration," that there was "no danger whatever of the man becoming a public charge," and that as a matter of "the simplest humanity and consistent with the laws of the land," Bosak should be admitted.

One of the inspectors who had already voted to exclude was moved to change his mind. The third, however, held firm for exclusion, which meant that the Secretary of the Department of Commerce and Labor, in which the Bureau of Immigration was located (after 1913, the Department of Labor), would have the final word. In his letter to the Secretary, the third inspector wrote that he wanted to "particularly call attention to Department letter 48,462," the recent memorandum that had emphasized the danger of admitting people with poor physique "whose offspring will reproduce, often in an exaggerated degree, the physical degeneracy of their parents." The commissioner at Ellis Island also strongly recommended exclusion, calling attention to the same memorandum. The secretary heeded their pointed references to the eugenic danger more than Cowen's appeal to "simple humanity" and ordered that Bosak be returned to the ongoing pogroms in Russia.[107]

Even in cases where the board unanimously decided against admission, immigrants had the right to appeal the decision to the Secretary of Commerce and Labor. Most did not, because they lacked either the skills, the resources, or the confidence to do so. Those who did appeal tended to be better educated and more articulate; in some cases they had the personal wherewithal to hire attorneys or assistance doing so from individuals or immigrant aid associations. The medical inspectors had one reasonably clear-cut responsibility: to detect and identify disease and defect. The duty of members of a Board of Special Inquiry was somewhat more ambiguous: when an immigrant with a medical certificate came before them, they were empowered to take factors other than physical condition into account, but which factors were legitimate was often a matter of debate. The commissioner general had greater freedom to make his recommendation based on whatever factors he thought the secretary might find relevant. The secretary, the final link in the chain, was restrained mainly by political considerations, and for many critics he was the weakest link. Sympathy for the immigrant's plight, pressure brought to bear by influential individuals or organizations, the state of public opinion, the inclinations of the president, and newspaper editorials all could come into account. Restrictionists as well as immigration officials often railed against what they saw as excessive leniency by the secretary.[108]

Chune Fruman, a twenty-year-old Russian Jew, arrived at Ellis Island in December 1906 and like Bosak was certified for poor physique, as well as curvature of the spine. At his hearing, he said that he was a blacksmith, that he was "a strong man and can work." The board unsurprisingly voted to exclude him as likely to become a public charge, but when his brother arrived from Chelsea, Massachusetts, to testify on his behalf, he was granted a second hearing. Also a blacksmith, with his own shop, a steady income, and a respectable bank balance, the brother said he would help Fruman until he had gotten on his feet, as would their other brother who lived in the same town. The board reaffirmed their decision to exclude. When the second brother came down from Chelsea to testify, the board reconvened for a third hearing. This time, however, Cowen was sitting on the board, and under his questioning the brother revealed that Fruman had fled military conscription in Russia. Recalling Fruman to testify, Cowen questioned him closely about his circumstances. Fruman produced his draft notice from the Russian government, which showed that he had undergone a physical examination and been "pronounced healthy and eligible for military service." Cowen asked what would happen to him if he was sent back. Because he had fled the country without a passport to avoid enlistment, he replied, "I would not be allowed to live for 24 hours."

Cowen again made a long statement before rendering his verdict. The Russian military was known to be exacting in its examination of recruits, he argued, and its affirmation of Fruman's fitness led him to "hesitate somewhat in excluding him on the certificate." He also revealed that "by strange coincidence" he had been on the same ship that brought Fruman from Hamburg, that severe weather had made it a difficult journey, and that he had observed Fruman "suffering very much indeed from sea sickness." That probably accounted for his poor appearance. Moreover, he believed Fruman's brothers would surely "look after him and see that he becomes quickly self-supporting." Nevertheless, given the medical certificate, he conceded that the case would have to be decided "by a higher authority," and hoped that in Washington "the matter may receive some special attention." Cowen's concession apparently did not entirely satisfy one of the other members of the board, who pressed the point that it was "the only function and duty of this Board to pass upon the eligibility of aliens as they appear before it," not to speculate or second-guess medical diagnoses. It was sufficient that "the alien has been examined by the properly authorized surgeons at this station and his physical condition pointed out, indicating, I think, that he is incapable of self support, and from every point of view he is an undesirable immigrant."

Fruman's file was forwarded to Washington, with a letter from the Ellis Island commissioner, Robert Watchorn, describing Fruman as a "man of very poor physical appearance" and "a most undesirable immigrant;" he therefore could not "agree with the inspector, who, for sentimental reasons, has importuned the admission of appellant." Fruman's brothers, however, apparently had connections or knew the right people to contact. Included in his file was an indignant letter from Congressman Ernest Roberts, whose district included Chelsea. Roberts argued in blunt terms that the "physical disabilities alleged are hardly such as would warrant the exclusion of any applicant" and that the board was "reading into the law" more than Congress intended. More important, Max Mitchell, of the Federation of Jewish Charities of Boston, made a personal appeal on Fruman's behalf to one Adolphus Simeon Solomons of K Street, Washington, D.C. A Washington resident since 1860, Solomons had recently turned eighty years old, and his October birthday party had been covered in detail by the *Washington Post.* Once an informal adviser to President Lincoln, he had been involved in the planning of presidential inaugurations ever since; he had been vice president under Clara Barton in the early years of the American Red Cross and was a member of the boards of numerous hospitals and charitable organizations; he had served in the Washington, D.C., House of Delegates and was currently director of the international Baron de Hirsch Fund, which was dedicated to, among other things, assisting Russian Jewish refugees. Commissioner General Sargent's report to the secretary briefly described the facts of the case before mentioning that he had "as well had a conference with Mr. Solomons, of this city." Sargent recommended that the usual rule be waived and Fruman admitted.[109] The secretary was persuaded.

Sargent, meanwhile, was worried about inconsistencies in the definition and application of the poor physique diagnosis at the various ports. The "Book of Instructions for the Medical Examination of Immigrants," which had been in use by the Public Health Service since 1903, defined it in terms similar to those used in Schereschewsky's memorandum but included other signs—such as "chicken breast," especially where tuberculosis was suspected but no bacillus found—suggesting a link between poor physique and communicable disease. Sargent wrote to George Stoner, head medical officer at Ellis Island, to ask his opinion of Schereschewsky's definition.[110] After two weeks, Stoner replied that the question had required lengthy discussions with his fellow medical officers, and "it was only to-day that an agreement was reached which may be regarded as a consensus of opinion." First, he reported, there was "no specific demonstrable disease sufficient to warrant a certificate" for poor physique. It was a catchall term, resorted to when no other diagnosis applied but "the defects present unfit the alien to earn a living at manual labor and lessen his power to resist disease." Stoner considered "the most important of these defects to be "lessened respiratory expansion," "deficient muscular development," "feeble circulation," and finally "lack of correlation between height and weight." Having proffered those details, he added that the condition "may be described briefly as defective physical development, a permanently faulty or reduced condition of the system, [and] a low degree of vitality," a description that appears considerably more capacious than the one he had just described. It may be that he was aware that the label was being used to exclude many immigrants who did not fit his more specific definition and did not want to be interpreted as criticizing those exclusions. By the end of his letter, however, Stoner more or less threw up his hands: "The foregoing is not intended as a complete or satisfactory definition. Indeed such definition would be difficult to frame. The difficulty arises from the fact that the term does not imply a clinical or pathological entity."[111]

In December 1906 the commissioner general threw his hands up as well, informing the Surgeon General that it "seems to the 'lay mind' of the Bureau that there is a wide divergence between the definition furnished by Dr. Schereschewsky and the description of the term given by Dr. Stoner." He had noticed that certificates for poor physique had "increased to such an extent that the said medical term has become perhaps one of the most important employed," yet no one seemed able to tell him exactly what it was. He closed with a plea that the physicians of the Public Health Service confer to arrive at a common definition, and "perhaps thus the matter could be placed in such shape that members of the boards of special inquiry, who of course with rare exceptions are strictly laymen, could pass more intelligently upon the cases under consideration."[112] Apparently no common definition was ever promulgated. Perhaps none was possible, for poor physique was never a medical diagnosis, but rather a nebulous description, promulgated largely by eugenicists, intended as an easily applied means of excluding degenerates.

Supporters of mandatory exclusion for poor physique found it difficult to get past the indeterminate vagueness of the term. A bill passed by the House of Representatives in 1907 would have made exclusion mandatory for a diagnosis of "low vitality or poor physique," regardless of economic fitness. The *New York Times* editorialized against the provision on the grounds that immigrants had a right to a hearing and should not be rejected on the basis of a single physician's judgment. "Public opinion," the *Times* argued, "will support amendments to the law that confirm or extend existing rules of exclusion based upon the fact of undesirability," but omitting hearings and the right to appeal risked "abuses of discretion." The Senate agreed, and the compromise legislation that finally emerged from the conference excluded instead those with a "mental or physical defect being of a nature which may affect the ability of such alien to earn a living." It fell short of the mandatory exclusion sought by restrictionists and left room for sympathy or politics on the part of the boards and the Secretary to intrude, but it had the political advantage of continuing to tie exclusion to an economic justification while at the same time considerably lowering the threshold for exclusion. Other immigrants would still be admitted unless "likely to become a public charge," but for those found to be defective, *likelihood* was no longer at issue. *Possibility* was the new standard—and not just the possibility of becoming a public charge, but even the possibility that a defect might have some unspecified effect on the ability to earn a living. Thus, beginning with the phrase "unable to take care of himself or herself" in 1882, the law had moved through "likely to become a public charge" in 1891 to persons whose defect "may affect" the ability to earn a living in 1907.[113]

The advocates of a physical test were gratified but not entirely satisfied by passage of the 1907 act. They continued to push for a more rigorous law that would exclude poor physical specimens regardless of any effect on employment prospects. Prescott Hall wrote to the Surgeon General the following year to request the annual total of poor physique cases excluded under the new law. He wanted him to know that "those who are watching the effects of the immigration law are much interested in the working of the so-called 'poor physique' clause, and are anxious to get definite and correct figures as to its operation," and he pointedly suggested that all future annual reports should have poor-physique cases enumerated in a separate table.[114] In his annual report of 1910 Commissioner General Keefe concluded that the new provision in the 1907 law had not been "by any means broad enough to reach all undesirables," and while he thought a literacy test would do some good, it would not be effective in "raising the general standard." Only a physical test would both improve the quality and reduce the quantity of immigrants. Keefe's primary concern was eugenic: while the immigrant's economic contribution in the present mattered, it was "more distinctly a matter of grave concern for the future."[115] In 1914 Commissioner General Anthony Caminetti repeated Keefe's call for a physical standard, what he called a "manhood test" to be modeled on the criteria used by the U.S. Army, in order to "accomplish simultaneously the two principal objects in the mind of all proponents of a stricter regulation and of a restriction of immigration: to wit, would improve the average quality of our immigration and correspondingly improve the physical qualifications of future generations within the country, and at the same time would work a material reduction in the number of immigrants admitted annually."[116]

Fears that the sheer scale of immigration would overwhelm the nation's ability to assimilate newcomers played a part in the restriction movement. But the intensity of restrictionist agitation did not closely correspond to the rise and fall of immigration. Moreover, the idea that the United States was experiencing a "flood of immigrants" was often contested. A conference on immigration hosted by the National Civic Federation in 1905, for example, pointed out that turn-of-the-century immigration was not large by historical standards if considered in relation to population.[117] A presentation at the 1912 meeting of the American Economic Association noted that, taking into account "birds of passage" (transatlantic migrant workers) in addition to permanent departures, net immigration relative to population from 1900 to 1910 was less than in any decade between 1840 and 1890, and roughly the same as the rate from 1860 to 1880.[118] The problem was not so much a flood of immigrants as it was that it was "a mighty tide of turbid and polluted water."[119]

Despite the argument that the impact of increased immigration ought to be gauged relative to overall population, the number of immigration stations and inspectors that Congress was willing to fund was still relatively inflexible. Increases in absolute numbers had a great impact on the efficacy of the inspection system. Even when provided with increases in staff, there really was a "flood" for the Immigration Bureau, and one that ebbed and flowed unpredictably enough that at times the ports were overstaffed and at others seriously understaffed. This made it impossible to effectively enforce laws that required careful inspection. Thus, for many advocates of reduced numbers, better inspection was the primary goal.[120] The real catastrophe, as the New York surgeon and medical researcher Arthur Fisk wrote, was that "an ever increasing horde of degenerates from all nations is entering our fair land, who are debasing the physical, mental and moral being of the nation."[121] The economist Irving Fisher, discussing in his first speech as the founding president of the American Eugenics Society in 1921 the potential "dysgenic" effects arising from modern war, hygiene, birth control, and immigration, emphasized that population growth or decline was in itself neither good nor bad: "The eugenist is interested in the quality of human beings rather than their quantity." The main immigration problem was that the United States had become "a dumping ground for relieving Europe of its burden of defectives, delinquents and dependents." He argued that "a discriminating exclusion must be eugenic," and he advocated "having aliens examined in their home countries for mental and other defects." Even when eugenicists advocated reduced immigration, their principal demand was a better means of selection, not reduced numbers.[122]

It became increasingly clear over the years, however, that selection via individual inspection could not accomplish the desired end without first reducing the numbers of immigrants. In 1914 Ward described the current laws as they existed "on paper" as "formidable." Over the previous three decades, "slowly, deliberately, carefully, this legislation was planned and grew up," and although it was an admirable achievement, "nevertheless, the experience of years has brought certain defects to light." The scale of immigration was overwhelming the inspection regime; as a result, the well-intentioned laws were "fail[ing] to accomplish their purposes."[123] It was simply too difficult, time-consuming, and expensive for inspectors to evaluate adequately the desirability of the multitudes of individuals that streamed past them. The assistant surgeon general of the Public Health Service told the House of Representatives in 1920 that "a complete medical examination requires at least an hour [and] one examiner could not handle more than 20 immigrants a day." Given the several thousand immigrants landing at Ellis Island most days, this would require greatly expanding facilities and more than tripling the current number of inspectors. Relying entirely on selection to restrict immigration was going to be expensive.[124]

Quotas

The problem of how to allow the desirables in while keeping the undesirables out was one that restrictionists and Congress grappled with for forty years.

The literacy test has been interpreted as a deliberate but veiled attempt to exclude certain races, but the fact that it was considered a blunt instrument at the time suggests otherwise. If it were aimed at racial groups, it would be a sharp instrument indeed, cutting out only selected individuals from within races. The problem was that it cut too broad a swath. Even if it were true that the illiterate were more likely to be unfit, it was clear that some unknown portion of the illiterate were in fact fit, and that many of the literate were equally unfit. In spite of these drawbacks, the test had the advantage of being less complicated, time-consuming, and expensive than the intensive individual inspection required to substantially reduce the number of defective types entering the country. When the literacy test finally came in 1917, however, its advocates were to be disappointed again. In the twenty-some years since it had first been proposed, public education and literacy in Europe had improved markedly, including among the classes considered most likely to be defective, and the test had little impact.

A quota system based on national origin was for many restrictionists a last resort, far from ideal but the only remaining practical alternative. Beginning with the Emergency Quota Act of 1921 and continuing with the National Origins Act of 1924, the means if not the ends of immigration restriction shifted. The 1921 act limited annual immigration from any one country to 3 percent of the number of immigrants from that country identified in the 1910 U.S. census. The 1924 Immigration Act reduced the number to 2 percent, substituted the 1890 census, and based the quota on proportions within the entire population rather than just within the foreign-born, thereby dramatically reducing immigration from southern and eastern Europe while granting a larger share to the British and other early immigrants. It furthermore barred all nonwhite persons "ineligible for citizenship," which effectively banned immigration from Asia.

In the years leading up to the quota acts, restriction advocates had often argued that certain European races were, in general, less desirable than others. For example, Alfred C. Reed, a Public Health Service physician and later professor of medicine at the University of California, contended that modern-day "Greeks offer a sad contrast to their ancient progenitors, as poor physical development is the rule among those who reach Ellis Island, and they have above their share of other defects." Jews, for their part, had a "predisposition to functional insanities," and overall "the proportion of defectives to total landed is greatest among the Syrians." Likewise undesirable were the "Magyars, Armenians and Turks." In short, "no one can stand at Ellis Island and see the physical and mental wrecks who are stopped there . . .

without becoming a firm believer in restriction." What this firm believer in restriction concluded, however, was that "strict enforcement of the present medical laws will automatically exclude these races to a sufficient extent, admitting the few who are fit."[125]

Defect and race were inextricably intertwined throughout the debates over immigration. A physician and specialist in medical sociology, Thomas Wray Grayson, warned about "the slow-witted Slav," the "poor physique" and "neurotic condition of our Jewish immigrants," and the "degenerate and psychopathic types, which are so conspicuous and numerous among the immigrants" of recent years.[126] The sociologist Frederick Bushee maintained that the "immigration of southern Italians brings a large superfluous population of hot-headed men," and that their rate of infant mortality "indicates small physical stability." Similarly, "the Irish have not that toughness, that power to resist disease," and their "physical instability is shown by the exceptionally large number of defectives among them."[127] The Public Health Service physician Thomas Salmon, who made the study of mental defects among immigrants his specialty, found "a remarkable number of mental defectives among Hebrews," followed closely by Italians and "other races of the new immigration," and detailed "psychoses" disproportionately diagnosed in each of these races.[128] The sociologist Edward Alsworth Ross, having observed immigrants disembarking at Ellis Island, wrote that "the physiognomy of certain groups unmistakably proclaims inferiority of type." Their bodies were noticeably inferior to those of earlier immigrants: "South Europeans run to low stature. A gang of Italian navvies . . . present, by their dwarfishness, a curious contrast to other people. The Portuguese, the Greeks, and the Syrians are, from our point of view, undersized. The Hebrew immigrants are very poor in physique . . . the polar opposite of our pioneer breed." One result was that "you can't make boy scouts out of the Jews. There's not a troop of them in all New York."[129]

Joseph G. Wilson, a senior Public Health Service physician, focused on one race in particular that was especially prone to insanity and feeblemindedness. Arguing that "the propagation of tainted stock" was the primary cause of mental defect, he warned of a "race who persistently refuse to practise the very doctrine which is essential to the preservation of a sound and healthy mentality. I refer to the Jews." Their "clannishness" had resulted in a "highly inbred and psychopathically inclined race . . . with a paranoid make-up." Only if they overcame their reluctance to marry outside the group would "Jews have it in their power to ultimately stamp out the feeble-minded and insane from among their race . . . It is all a question of eugenics." The danger to the United States was evident: "In 1907 nearly one third of those certified at Ellis Island as mentally defective were of this race, although they did not average over 14 per cent, of the total number of arrivals."[130]

By the 1910s race was practically defined by defect in discussions of restricting European immigration. Defect itself was not defined by race, however. A defective person of any race was unwelcome. Moreover, most restrictionists believed that excluding defective individuals would be the best means of reducing immigration from the inferior races. Harry Laughlin, for example, has been described primarily as a proponent of race-based immigration laws and an important influence on the passage of the quota acts. He was appointed Expert Eugenics Agent to the House of Representatives Committee on Immigration in 1920 and provided numerous reports and testimony while the committee was working on various quota acts over the course of the decade. In his first appearance before the committee in 1920, he opened by stating that the "character of a nation is determined primarily by its racial qualities; that is, by the hereditary physical, mental, and moral or temperamental traits of its people." This has been much quoted but gives a misleading impression of Laughlin's main preoccupations. In his testimony he did not advocate any sort of quota system based on race. Instead, he proposed careful examination of prospective immigrants "in their home towns, because that is the only place where one can get eugenical facts." Those essential facts did not come from knowing the race of the immigrant, but from examining his or her hereditary background. Restriction had to proceed at the level of the family and the individual, he believed, not the race or nation, in order to ensure "the possession in the prospective immigrant and in his family stock of such physical, mental, and moral qualities as the American people desire to be possessed inherently by its future citizenry."[131]

As examples of the dangers that poor family stock posed to the nation, he cited the well-known studies of the Juke, Kallikak, and Ishmael families—all of whom traced their ancestry back to colonial America and before that to England—as well as the "excessively large slum district" in Sydney, "populated to a considerable extent by the descendants of the Botany Bayers deported from England." From these examples, Laughlin concluded, the "lesson is that immigrants should be examined, and the family stock should be investigated, lest we admit more degenerate 'blood.'" The literacy test was beside the point and a poor measure of an immigrant's worth, because "if his children can be taught to read we would consider him good stuff." The threat came from what he called "social inadequates," otherwise known as the "defective, dependent, and delinquent classes," among whom he included the insane, the feebleminded, the blind, the deaf, the deformed, the crippled, paupers, and criminals. Above all, "the great problem" was the "problem of the moron," for while idiots and imbeciles were easily detected, "a moron can slip through the immigration sieve, as it exists to-day, pretty easily." Worse yet, "some of those moron girls physically are well developed" and "highly fertile sexually." To be sure, Laughlin testified that morons were more common among recent immigrants, along with "practically all other types of the socially inadequate," but this was "not so much a matter of nationality—that is, northern European blood against southern European blood, as it is skimmed milk versus cream in each of the countries sending us immigrants." Indeed, it was "doubtful whether there is a single country in the world that does not have many families so splendidly endowed by nature that they would not make excellent and desirable additions to our citizenry." His argument was not based on blanket characterizations of particular races. Instead, he made a subtly but crucially different point: the "lower or less progressive races" supplied a greater percentage of defectives, and the worst of each nation was increasingly migrating thanks to the ease of transportation. These two factors, for Laughlin, both demonstrated that a problem existed and explained why it had come about. While he called upon racist assumptions to explain the problem and its origins, race was primarily a shorthand way to describe relative risk of defect. The crux was that "at present, not inferior nationalities but inferior individual family stocks are tending to deteriorate our national characteristics," and it was the "failure to sort immigrants on the basis of natural worth" that constituted a "very serious national menace."[132]

In Congressional testimony in 1922, after racial quotas had been already implemented, Laughlin presented detailed charts purporting to show the prevalence of various defects among European immigrants living in the United States. Again, these charts have been much cited and reproduced, but Laughlin cautioned the committee that they represented "not a direct measure of relative racial values" and were useful only as a "measure of the degree to which each racial and nativity group must be culled or sifted . . . and later, if this fails, in deporting degenerate individuals and families of whatsoever nationality." His research would help immigration officials to predict what sorts of "degeneracies and hereditary handicaps," which be believed to be "inherent in the blood," they could expect to find in individuals from particular racial groups, as well as what percentage of individuals of each race they should expect to reject on grounds of unfitness. The hierarchy of European races and their compatibility with the so-called American race became increasingly important elements in Laughlin's thinking throughout the 1920s and 1930s, yet they never supplanted defect as his central concern, nor did he ever cease to advocate individual inspection abroad as the best solution to the problem.[133]

Although medical inspection abroad was instituted in 1924, the kind of extensive investigation that Laughlin favored was deemed unworkable, in part because foreign governments viewed it as violation of sovereignty.[134] In the quota laws, however, restrictionists found a workable and relatively efficient formula for substantially restricting immigration. Like the literacy requirement, passage of the quota laws was not a departure but a continuation of the search for an effective means of excluding defective people. The quota acts did not supersede existing laws but complemented them. The earlier laws remained in place, and selection by physical inspection continued. Congress added to those laws a system that selected immigrants of desirable nationality and race and deselected those most likely to produce defective strains. The continued exclusion of defectives was understood to affect immigrants from inferior nations disproportionately, while quotas on immigration from those nations likewise reduced the numbers of defective immigrants that had to be dealt with by inspectors. The laws continued to be simultaneously selective and restrictive. Quotas were a continuation of a policy of restriction via selection, only by other means.

Once the quota laws were in place, what did restrictionists have to say about them? Writing in 1924, Ward noted the importance of good "racial stock" but emphasized that the main benefit of quotas was that, "until we had set some sort of numerical limitation on immigration, better selection—mental, physical or economic—was impossible." Earlier laws had attempted to bar those who should "be excluded because eugenically undesirable," but in spite of great effort "thousands of mentally and physically unsound aliens landed . . . because the hurried medical examination here can not detect the disabilities" that should have barred them. Quotas were an important step, Ward wrote, but did not adequately address his chief concern, and he returned to the theme that had animated his political efforts for years: "It now becomes our duty, both to ourselves and to coming generations of Americans of whatever racial origin, to set higher mental and physical standards for all our future immigrants."[135]

Glenn Frank, editor-in-chief of the *Century Magazine* and later president of the University of Wisconsin, wrote in opposition to the proposed quota legislation in 1924. He did "not dispute the fact that races have differing endowments of blood and culture," nor did he look with favor on "the promiscuous interbreeding of races." Furthermore, he agreed that "the hour for very severe restriction on immigration has come." His objection to the law was that it was "arbitrary and mechanical," a "lazy method of rigid racial classification and mathematical percentages," ineffectively targeted compared to "a truly scientific regulation of immigration" to "select our immigrants upon the basis of their individual fitness." What was needed was a dedicated effort to improve the technology of selection. As "mental tests are perfected," as a "method for getting the family background of prospective immigrants" was developed, and as scientists "explore[d] the possibilities of the science of human measurement," he was confident that a "truly selective immigration policy can come into effective operation." Scientific regulation would more effectively screen out the unfit than "arbitrary racial classification," while at the same time favoring superior races: "we Nordics . . . need have no fear that we shall not fare well in a scientific selective process."[136]

The Report of the Eugenics Committee of the United States Committee on Selective Immigration supported the more stringent quotas instituted in 1924, but mainly because they would "greatly reduce the number of immigrants of the lower grades of intelligence, and of immigrants who are making excessive contribution to our feebleminded, insane, criminal, and other socially inadequate classes."[137] Laughlin, for his part, continued to believe that the country had yet to establish an entirely satisfactory policy from a eugenic point of view. In 1924, he remained convinced that "our future laws, if the country is to be protected against inferior immigrants and is to select and welcome superior strains, should provide by statute for the determination of individual and hereditary qualities by requiring modern pedigree examination in the home territories of the would-be immigrant." In short, the answer was "to perfect the principle of selective immigration based upon high family stock standards."[138]

In 1925 the Secretary of Labor, James J. Davis, looked back over the history of immigration policy and reflected on the present. As he saw it, the main reason that Americans had chosen to gradually narrow their gates to immigrants over the previous four decades was that their country had "become an asylum for the alien insane, defective, and degenerate." The failures of the inspection system had led the nation to finally resort to quotas. Inspection alone had proved incapable of coping with the "great wholesale pressure of immigration which would completely swamp us." The best argument in favor of quotas, he believed, was that they slowed immigration to the point that careful and rigorous selection was possible. If that had been possible to accomplish without quotas, the "retail method of selection" alone would be "the fairest test of fitness" and "the better way of selecting our guests and future citizens." He believed that "most civilized races contribute good, sound strains of family and individuals." Race had never been the central issue, he insisted, but it had to be acknowledged that certain "races are spotted with defective, degenerate, and inferior lines and stocks."[139]

The fundamental role of defect in the history of American immigration law has received less critical inquiry than its prominence in the laws warrants. Defect—or disability—was the central issue in restriction debates and legislation into the 1920s, and most of the other categories of exclusion were understood, at least in part, in terms of defect: an undesirable race was one in which defects proliferated; the diseased might have degenerate constitutions that made them susceptible; criminals and the otherwise immoral were feebleminded or "moral imbeciles;" deviant sexuality was a mark of mental defect; political radicals were mentally unbalanced; and poverty was symptomatic of inborn psychopathic inferiority. Even the capacity to assimilate in American society and adopt democratic norms was said to depend upon certain heritable mental and moral qualities that some possessed and others lacked.[140]

Understanding early immigration law as unfolding in two distinct phases, selective and restrictive, obscures the significance of defect in this history. The laws during this period are better understood as forming a cohesive whole, a decades-long effort to find an effective method of excluding immigrants seen as defective. Lawmakers at first tried to restrict immigration via increasingly stringent admission standards and inspection procedures. They assumed this would reduce the numbers not only of defective individuals, but also of those races understood to be disproportionately prone to defect. When screening for defects proved incapable of checking the flow of undesirable immigrants, legislators finally turned to a system of quotas based on national origin. This was seen not as a departure from but rather a complement to the previous acts. The inspection and exclusion of defective individuals was expected to reduce the numbers of immigrants from defect-prone races; conversely, national quotas were intended to further reduce the number of defective individuals. The overarching goal did not change. In immigration law, as in other realms, the issues of race and defect were deeply intertwined.[141]

The history of immigration policy has focused primarily on the analytic categories of race and ethnicity, followed by class and economic interest. To the extent that eugenics has been included in the story, it has been to decry its invidious application to ethnoracial groups. This focus may well be a result of the influence of the larger field of immigration history, which has developed largely as the study of ethnic groups, an emphasis reflected in the title of the leading journal in the field, the *Journal of American Ethnic History*. It may also have to do with the fact that class, race, and ethnicity deal with recognized and politically salient groups. Although people with disabilities also formed groups, networks, and organizations built around shared values and common interests, they were neither large nor widely known and were typically organized around particular disabilities.[142] There were no calls for defectives of the world to unite, no pan-disability associations. In his 1922 polemic against the eugenics movement, G. K. Chesterton lamented that "there is no trade union of defective children," and that therefore most expressions of opposition to eugenics amounted to "protests so ineffectual about wrongs so individual." Chesterton feared that this same dynamic would make the injuries inflicted on disabled people less visible in history, "trivial tragedies that will fade faster and faster in the flux of time."[143]

Defectives did form a large and powerful group, however, in the imagination of eugenically minded Americans. There the defective population assumed frightening and ever-growing proportions, even while uncounted others from abroad clamored to add to it.[144] The exclusion of "the defective" was a matter not merely of rational economic considerations but, in the context of an all-pervasive ethic of competition, of national survival and the preservation of the American race. The heritability of virtually every personal characteristic, the mutability of defects across generations, and the deplorable fecundity of degenerates evoked the specter of racial declension. This was the meaning and the import of "selective" immigration in a eugenic nation.

2

Handicapped

At the turn of the twentieth century, the meaning of time was rapidly changing. On the one hand, the spread and popularization of discoveries in geology and evolutionary biology were radically altering the meaning of historical time. On the other, the expanding industrial economy was greatly accelerating longer-term changes in perceptions of everyday time. These twin revolutions in thought brought into being new ideas about disability and a new vocabulary with which to talk about it. Terms such as "handicapped," "retarded," "abnormal," "inefficient," "defective," and "degenerate" came into common use, each of them implicitly or explicitly rooted in the new ways of understanding of time. When immigration restrictionists talked about "undesirable" immigrants, they understood that term in the context of the new significance and meaning of time.[1]

Affliction in a Designed World

Until the late nineteenth century, "infirmity" and "affliction" were the terms most commonly used to refer to what we now call disability, though both were comparatively more capacious terms. "Infirmity" denoted weakness, frailty, or debility from any cause, the "infirmities of age," "an infirmity of will," or "an infirmity of purpose" being commonly heard phrases. "Affliction" was applied to almost any difficult or unfortunate circumstance. Disabled people shared it with the poor and the sick, widows and widowers, parents whose children died young, and anyone who suffered nearly any unpleasantness, from tragedy to inconvenience. Thomas Jefferson on various occasions wrote that he was "afflicted" with a headache, that the "affliction of the people for want of arms is great," that a tiresome colleague was "afflicted with the morbid rage of debate," and that the death of Benjamin Franklin was an "affliction" to those who knew him. Herman Melville wrote in *Moby-Dick* of an aged whale "afflicted with the jaundice, or some other infirmity" and of marital troubles as "domestic afflictions."[2]

The experience of infirmity and affliction was common and universal. When a significant or long-lasting condition was referred to, both words (but especially "affliction") in American popular usage often carried the connotation of imposition by an omnipotent god, an aspect of a larger design, and a spiritual burden to be borne or lesson to be learned. Didactic stories for children often featured afflicted children—for example, one was about a girl who "was crooked and rather lame" and whose "affliction was a sore trial to her, for she had not yet learned that it was sent by a wise and loving Father, who knows what is for our real good."[3] In the fiction of the 1840s, Penny Richards has shown, feebleminded children were often depicted as simultaneously an affliction and a gift to their families, mothers especially, who were called upon to make sacrifices but in return gained wisdom, patience, and love.[4] Affliction also taught harsh lessons. A teacher at mid-century wondered "whether sin had brought into our world any heavier affliction" than deafness. Another advised a man on the importance of reconciling himself "to the will of Providence which made him deaf." Whether seen as just deserts for misdeeds or the design of an inscrutable will, disability was commonly said to be the "decree of Providence" and the "condition in which God has placed them"—at least "until the time arrives when human imperfections will be done away with."[5]

Ministers constructed sermons on the meanings to be found in affliction. In 1851 the Reverend James Caughey, addressing those who lament the "great mystery, why the Lord suffers his children to be afflicted," gave the customary assurance that the "reasons are unquestionably wise, though we may not be able to find them out in this world." Having made the necessary nod toward the utter impotence of human comprehension in the face of God's inscrutability, he nevertheless proceeded to explain affliction's purpose. Alchemists who "turn baser metal into gold," he began, "have always acted first upon the Latin motto, *Abstractio terrestrietatis a materia*, 'The abstraction or drawing away of earthliness from the matter of their metal.'" God, likewise, having as raw material "our base souls," places them into the "furnace of affliction," where they are purified, their earthliness abstracted and "turned into fine gold."[6]

The concept of affliction inhabited a larger belief system in which the world was a product of conscious design, essentially unchanging and infused with purpose. Prior to the ascendance of scientific naturalism in the late nineteenth century, most Americans imagined a relatively short arc of history of only a few thousand years, from creation to imminent and preordained end. Christians debated whether humans possessed free will, but not to the extent of challenging the notion of design. The question came down to whether they were predestined to follow a path laid for them or could choose to deviate from it, thereby alienating themselves from the Creator's plan and slowing humanity's progress toward its ultimate goal. In either case, a path had been laid. For those Protestants who rejected the doctrine of predestination, the human project on earth was to seek to understand the creator's intent as revealed in nature and to strive to live in accord with it, thereby taking part in its development toward its intended state.

The bodies of living things were similarly explained in terms of purpose and intent, from the perfection of birds' wings for particular kinds of flight to the opposable thumbs of humans for holding tools and the tails of monkeys for swinging through trees. The best-known version of this story was found in William Paley's immensely popular book *Natural Theology*, published in 1802 and used in high school and college classrooms well past mid-century. Paley wrote that anyone who found a watch on the ground would immediately understand that it was "formed and put together for a purpose"; the obvious inference was "that the watch must have had a maker . . . who formed it for the purpose which we find it actually to answer; who comprehended its construction, and designed its use." If this were true of a mere watch, how much more true of the intricate design of living things. Paley led his readers on a walk through a world in which evidence of a maker was found everywhere, in the humps of camels, the fangs of snakes, and the snouts of pigs. His assumption, shared by most people in the West, was that since God cannot be capricious, understanding the world meant asking what purposes God had in mind while constructing it.[7] Richard Owen, the world's leading pre-Darwinian naturalist, wrote in 1848 that every species had abilities suited to their needs "in a way that indicates superior design, intelligence and foresight," which an observer "must ascribe to the Sovereign of the universe."[8] Darwin himself subscribed to the design theory before the evidence he gathered on his voyage on the *Beagle* rendered it untenable; he later described it as the "old argument of design in nature, as given by Paley, which formerly seemed to me so conclusive."[9]

The design idea today is typically associated with biblical literalism, but this was not necessarily the case in the nineteenth century. Biblical literalism was one strain in Protestant thinking but was not as widespread as it would later become. The new discoveries in geology were widely accepted, and most Christians who held a creationist understanding of the world had no problem with the idea that the earth was millions, if not billions, of years old. Various ideas circulated to reconcile the biblical stories with modern science: the day/age theory, which explained the seven days of creation as a metaphor for seven ages, for example, or the gap theory, which suggested that the earth itself may have been created in the distant past, followed by a great gap of geological time and then Eden as described in Genesis. Relatively few Christians thought science incompatible with their beliefs.

In schools across the country, children read geography textbooks that explained the planet in terms of design, invariably introducing the object of study as the "temporary dwelling place of mankind" or the "earth which was made to be our home." The geographical features that the books described had been fashioned to accommodate human life, and "to that end, were created the land, with its mountains and plains; the water, with its mighty ocean and running brooks." Nothing was superfluous or out of place; nothing, for that matter simply *was*—everything was for a reason. Because the earth "could not be the abode of man" without them, "two indispensable agents are provided—the sun and atmosphere." Because those who lived in the "torrid regions of the earth require the greatest amount of rain, there are the loftiest mountains, which act as huge condensers of the clouds." Alternatively, because breezes from the mountaintops cooled inhabitants of the valleys, mountains were located in hot areas "for the same reason that you put a piece of ice into a pitcher of water in summer, rather than in winter." In any case, had "the highest chains been situated in the polar climes, they would have been encased from base to summit with glaciers and snow, and thus rendered uninhabitable; but placed as they are, a vast extent of surface has been rescued from desolation." On the subject of oceans the authors evinced some defensiveness, given that a planet specially designed for a terrestrial species had more water than land, but they chided those who "fail to see that the ocean, which to the thoughtless appears as a great waste, is vast in its benefits." For example, while the "ocean covers four-fifths of the Tropics," this proportion was necessary to supply "the rainfall of the Temperate regions," and if the "land of the Tropics were greatly increased, heat would so accumulate that all animal and vegetable life there would be destroyed." There was also the "interesting fact that the best fish are found in the cold currents, near the coasts," where they were more conveniently caught. Concerning water in colder climes, why did it expand when frozen, unlike other substances? Because, if it contracted, "the freezing particles being heaviest, would sink to the bottom," resulting in bodies of water gradually freezing from the bottom up "which no summer's sun would have power to melt." This "remarkable exception to a law otherwise universal is, therefore, a means of preserving, in cold climates, the liquid form of this element." Finally, in the best example of this form of reasoning, "As the earth is round, only half of it can be lighted at once. In order that both sides may be lighted, the Creator has caused the earth to rotate."[10]

The existence of afflictions in the world was often hard to explain, but its place in a larger design was nevertheless assumed. In an address before the Connecticut legislature in 1818, Laurent Clerc, cofounder of the first school for the deaf in the United States, explained the existence of disability in a way that could make sense only in a designed and static world. Having stated the commonplace belief that "every creature, every work of God is admirably well made," he acknowledged as well the plain fact that "everything is variable and inconstant," and some things "we do not find right." When people look around them, they see that "here the ground is flat; there it is hilly and mountainous; in other places, it is sandy; in others it is barren." They find "trees high or low, large or small, upright or crooked, fruitful or unfruitful." There are animals that are "harmless or ferocious, useful or useless, pleasing or hideous"; some of them suffer from "faults in their organization." As for humanity, "our intellectual faculties as well as our corporeal organization" are likewise prone to imperfections. Intellectual shortcomings might be ameliorated by education, "but nothing can correct the infirmities of the bodily organization, such as deafness, blindness, lameness, palsy, crookedness, ugliness." In spite of this, he insisted, "everything is well made," for much as people might prefer trees to be tall, upright, and fruitful and animals to be harmless, useful, and pleasing, the world was made not to satisfy superficial human preferences but for purposes beyond their ken. "Why then are we Deaf and Dumb?" Clerc asked. His response: "I do not know, as you do not know why there are infirmities in your bodies, nor why there are among the human kind, white, black, red and yellow men." Because these were aspects of a larger design, and the creator could not be capricious, he concluded that the deaf "cannot but thank God for having made us Deaf and Dumb, hoping that in the future world, the reason of this may be explained to us all."[11]

Handicapped in the Race for Life

By the 1890s, however, the idea of affliction and the worldview that sustained it began to give way to the very different idea of "handicap." The word itself originated in a popular game of chance in seventeenth-century England known as "hand-in-cap" or "hand-i'-cap." The rules were arcane and accounts of them vary, but the game involved two players who each put up some item of value, an umpire who estimated the difference in value between the items, and, of course, a hat. Beyond that it gets rather foggy. Players apparently placed their hands in the cap clutching money, the umpire pronounced his verdict, and the hands were withdrawn either empty or still holding the money, which signified acceptance or rejection of the umpire's judgment. The principal goal of the game was to compare items of unequal value and estimate the difference between them. This essential idea later, in the eighteenth century, migrated to horse racing: in a "handicap race," an umpire, or "handicapper," compared horses, estimated the differences in speed among them, and determined how much weight the faster ones had to carry in order to even the odds.[12] Wagers would be placed based in part on judgments concerning the handicap.

Over the course of the nineteenth century, the idea of a handicap was generalized to other kinds of matches in which the superior competitor was purposely disadvantaged. Handicapping was applied to foot races, yacht races, rowing, billiards, card games, badminton, croquet, even pigeon shooting. Eventually the idea drifted out of the sports orbit to inhabit, as a metaphor, nearly every competitive arena, as when the *New York Times* in 1874 ventured that southern Republicans had "no desire to handicap the party with any extreme doctrines on civil rights." A newspaper advertisement in 1897 warned, "Don't handicap your appearance by wearing a shabby looking out-of-date hat." In 1908 an *Atlantic Monthly* writer ventured that poor "economic and social efficiency" might "constitute serious handicaps in the matrimonial race."[13] Not surprisingly, in a competitive age, disability also began to be spoken about with the language of handicap.

This new language of competition and its diffusion into so many spheres of life was rooted in the convergence of two major changes in the ways that Americans thought and talked about time. One had to do with historical time and was spurred by the advent of evolutionary science, the other with the increasing competitiveness of everyday life amidst the rise of a market economy. One of Darwin's insights that led to the theory of natural selection was that all living things competed for scarce resources. Those with some advantage would be more likely to survive, those less advantaged less likely. Life at its core was competition. For several decades following the publication of *The Origin of Species* in 1859, no particular theory of evolution gained dominance in the United States, but the basic tenets of evolutionary science quickly found widespread acceptance. Evolutionary metaphors, analogies, and explanations soon became ubiquitous across practically every field of scholarship as well as in the broader popular culture.[14] As Richard Hofstadter wrote of "social Darwinism," too many found it "easy to argue by analogy from natural selection of fitter organisms to social selection of fitter men. . . . The competitive order was now supplied with a cosmic rationale. Competition was glorious."[15] Scientific naturalism in general, and evolutionary science in particular, radically altered how Americans defined themselves and their world. In concert with earlier discoveries in physics and geology, evolutionary science vastly expanded the dimensions of time. Past and future no longer appeared as an unfolding plan but rather as a series of contingencies, open-ended and indeterminate, both exhilarating and terrifying. Nearly any future became thinkable. The hope placed in progress had as its constant companion the fear of decline, and new branches of literature sprang up to explore myriad imagined futures, some bright and others despairing.

Geography schoolbooks, with their cozy vision of a world designed and harmonious, "made to be our home," by the late nineteenth century depicted vast impersonal forces acting over huge expanses of time. Whereas Brocklesby's *Elements of Physical Geography* told readers in 1868 that the "physical phenomena of the world reveal in their harmonious action a unity of plan and purpose, and display in an infinite variety of ways the Power, Wisdom, and Goodness of the Almighty Designer," the 1901 *Rand-McNally Grammar School Geography* offered instead "a connected chain of causes and results, every link of which presents a problem to stimulate investigation and awaken rational thought."[16] Mountains, those great air conditioners and rainmakers devised for the benefit of the torrid regions, became mere "wrinkles or folds, made by the shrinking of the earth's crust while it was cooling," while the carefully planned oceans became merely "wide hollows or valleys in which the waters of the sea collect."[17] The world existed in a constant state of flux, one thing causing another, changing into another, passing by insensible gradations into every other thing, and what determined the present form of living things was a great chain of competitive encounters, life vying with life, competing for a place in the world.

How Americans experienced time day to day was being similarly destabilized. A culture that celebrated individualism, competition, and achievement was emerging from the rapid spread of a market economy, industrialization, and concomitant urbanization. New terms emerging in popular speech, such as "individualism" and "self-made man," reflected the change. "Failure" had previously been a term to describe a business venture that went bust, but with the rise of a market economy and a growing faith in the virtues of individual competition, it expanded its purview from an unfortunate incident to a personal identity and state of being.[18] With the growth of nationalism alongside mounting international competition added to the calculation, stasis seemed no option, for to stand still was to fall behind.

That the pace of life was increasing rapidly was obvious not just in hindsight but to people at the time and often discussed, with approval or alarm.

William R. Greg wrote in 1875 that "beyond doubt, the most salient characteristic of life in this latter portion of the 19th century is its SPEED,—what we may call its hurry, the rate at which we move, the high-pressure at which we work." Greg was British, but as an editorial in the *New York Times* that same year noted, "Mr. Greg's remarks have even more point here than in his own country." Josiah Strong, in his popular 1891 book *Our Country*, wrote that "it must not be forgotten that the pulse and the pace of the world have been marvelously quickened during the nineteenth century." Indeed, the American "finds life growing ever more intense and time more potent. . . . There is a tremendous rush of events which is startling, even in the nineteenth century." Strong segued easily between the quickening pace of daily life and the gathering speed of progress, suggesting the close relationship between the two. That "rush of events" could refer to everyday experience, or to his observation that the "western world in its progress is gathering momentum like a falling body." In either case, he wondered, "can anyone doubt that the result of this competition of races will be the survival of the fittest?"[19]

In his 1901 story "The New Accelerator" H. G. Wells had a scientist concoct "an all-round nervous stimulant to bring languid people up to the stresses of these pushful days," something to confer "the power to think twice as fast, move twice as quickly, do twice as much work." It was an elixir to give a competitive edge in an accelerating world, to make "you go two—or even three—to everybody else's one." The fantasy was a pleasant escape from the certain knowledge that no matter how fast any individual managed to go, no advantage could last long, as everyone and everything else was constantly accelerating. The felt necessity to use time more efficiently became pervasive. Traditional standards for the pace of work were being quickly discarded as, in an intensely competitive free-market economy, the essential need was to be as fast as or faster than the next business or worker.[20]

The "gospel of efficiency" permeated nearly all fields of endeavor, from businesses to schools, hospitals to prisons. The historian Samuel Haber once observed that "efficient and good came closer to meaning the same thing in these years than in any other period in American history."[21] Frederick Winslow Taylor's 1911 treatise *The Principles of Scientific Management* urged that the same principles that applied to the efficient operation of the factory be "applied with equal force to all social activities: to the management of our homes . . . our churches, our philanthropic institutions, our universities." According to Michael Knoll, "politicians, businessmen, and scientists embraced the 'gospel of efficiency'" and "discussed untiringly how they could improve the 'national,' 'industrial,' and 'scientific efficiency' of their country, company, or college."[22] A teacher of deaf students was asked, in 1907, what should replace the inefficient teaching methods of the past. "The answer may be given in a word," he answered—"the factory, the factory principle and system in everything." Another teacher told his colleagues at a 1914 convention that "in this modern day everybody and every movement is coming to be measured by and for efficiency."[23]

A system of production increasingly based on wage labor outside of the home inexorably drew people with disabilities along with everyone else into a competitive labor market (children and the elderly as well, until compulsory education, child labor laws, and social security insurance took them out of the market). Earlier in the century, with limited exchange and home-based production, the key to survival had been for every family member, young, elderly, or disabled, to contribute what he or she could at whatever pace was manageable. With the growing dominance of wage labor, in tandem with the social value placed on speed and efficiency, many found themselves unable to contribute to their own support, whether because of functional impairments, discrimination, or some combination thereof. More and more they were described as "burdens" on the family as well as the community and nation, all of which were engaged in various competitive spheres. Any handicap was cause for not only individual but also familial and social apprehension.[24]

At first, "handicapped" as a descriptor of disabled people came embedded in explanatory phrases, such as "handicapped in the race for life," or "by this defect handicapped in the struggle for existence."[25] Gradually, however, handicap came to be associated with disability more than with anything else. In 1890, when an obituary eulogized the subject's "life work done in spite of the handicap of physical infirmity," it was still necessary to specify that the handicap resulted from infirmity, and a newspaper article in 1891 titled "Handicapped Children" had to explain that it referred to children who are "handicapped by some physical ailment or disability."[26] Other factors, such as racial identity, might also be described as a handicap, as when a judge concluded that the plaintiffs before him were "of sufficient Indian blood [as] to substantially handicap them in the struggle for existence."[27] But by the early twentieth century, to say "the handicapped" was sufficient, without qualification, to be understood as speaking of disabled people. In 1911 the radical writer Randolph Bourne could title his essay in the *Atlantic* "The Handicapped—By One of Them" without ambiguity, and the social reformer Lillian Wald in 1915 could head a book chapter simply "The Handicapped Child" without explanation. The efficiency experts Frank and Lillian Gilbreth noted in their 1920 book *Motion Study for the Handicapped* (which itself needed no subtitle) "the universal use of the word 'handicapped' today instead of the 'crippled.'"[28]

FIGURE 5. The phrase "handicapped in the race for life" was so ubiquitous that it could appear even in an advertisement for eyeglasses. Museum of Vision, American Optical Company collection, Accession no. AOC.2.001.0024, c1930. (Courtesy of the Museum of Vision and The American Academy of Ophthalmology. All rights reserved.)

The idea that life was a race explains much of the impetus behind the demand for an increasingly strict immigration policy. There was not one but two races, two struggles, in which disability was a handicap: the economic race for life in which an individual might succeed or fail, and the evolutionary competition by which races and nations would rise or fall. The economic and the eugenic rationales for exclusion were not, however, distinct issues, as they are often treated, but closely linked. One purpose of the laws was no doubt to eliminate those thought more likely to fail individually in the economic race. For example, Boston immigration officials refused entry to Leopold Perelman in 1905 because as a deaf man he could not "compete with an ordinary workman in this country."[29] That same year the commissioner general advised the exclusion of Chave Unger because her deafness would "greatly handicap her."[30] As if handicapping a race, immigration officials sometimes gauged degrees of handicap with implausible precision, as when the inspecting physician described Bernhard Mydland's curved spine in 1910: while "the deformity is 50%, I cannot say the physical handicap would be that much–I should say 25%."[31] The pioneering social worker Edith Abbott wrote in 1924 that an immigrant's feeblemindedness "will decidedly handicap him among his fellows in the struggle for existence."[32]

At the same time, the greater problem to which these individual failures was linked was the long-term effect on the nation's own struggle for existence. The New York Supreme Court Justice Norman S. Dike, for example, worried that the new immigrants were "adding to that appalling number of our inhabitants who handicap us by reason of their mental and physical disabilities."[33] In a rousing speech to the American Association for the Study of the Feeble-Minded in 1913, the governor of Michigan declared that "so long as we care less about the kind of human beings produced than we do about the kind of livestock, we shall be handicapped in our progress." Americans, he argued, understood that when "they talk about fine horses, fine cattle, fine sheep, fine fruit, they are standing on a commercial foundation," and if only discussions of human quality could be placed on the same sturdy foundation, the country would be willing to support "more stringent marriage laws [and] immigration laws."[34] A year later the eugenics section of the American Breeders' Association warned that the alarming number of Americans who carried an "inherent defect" were an "industrial and social handicap," a "danger to the national and racial life," and a "drag on society."[35] In 1922 the distinguished historian William Roscoe Thayer asked, "Why should the United States accept the handicap of inferior candidates of whatever race?"[36]

Retarded

Although physicians and educators had begun in the eighteenth century to reframe intellectual disability as a condition caused by some form of deprivation, most laypeople and many professionals still discussed intellectual disability in moral and religious terms well into the nineteenth century. The ideas of Samuel Gridley Howe, director of the first state institution for the feebleminded, and other mid-century experts were complex and influenced by psychology and phrenology, but prevention and cure still often came down to following the laws of nature established by God. Howe attributed the existence of feeblemindedness to parents' moral lapses—"where there was so much suffering," he asserted, "there must have been sin"—and whether meant to provide a lesson or punishment, it was an aspect of intentional design.[37]

Religious explanations declined in professional circles over the course of the century, and evolutionary ones began their ascent in the 1860s, as when John Langdon Down, medical superintendent of the Earlswood Asylum for Idiots in southeast England, identified the syndrome later given his name as "mongolian imbecility," or "mongolism." The shape of the head, facial features, and intellect of some of the residents in his asylum suggested to him a biological reversion by Caucasians to the Mongol racial type. He thought mongolism to be the most common form, but Down created a complete "ethnological classification of idiots" that described Mongolian, Ethiopian, American Indian, and Malay types of the feebleminded. Many experts resisted Down's theory, some arguing that because mongolism occurred also among African Americans, members of a race presumably lower than Mongolians on the evolutionary ladder, it could not be a case of reversion; others were skeptical for a variety of other reasons. Nevertheless, the idea persisted.[38]

Carlisle P. Knight, a Public Health Service physician at Ellis Island, writing on "detection of the mentally defective" in 1913, advised his colleagues that the "various ethnological types" of mental defectives were "easily discerned: the dark skin, the curly hair and thick lips of the Ethiopian, the prominent and high cheek-bones and deep orbits of the American Indian and the straight coarse hair and peculiar cast of countenance of the Mongolian."[39] In the 1920s, elaborating on Down's theory by incorporating Mendelian genetics, the British physician and writer Francis Graham Crookshank suggested that mongolism was likely caused by a recessive "unit character" that persisted in Western populations and occasionally resurfaced, "coarsely and brutally displayed and accentuated in certain idiots and imbeciles."[40] In the United States, Charles Davenport was still advancing a theory of racial classification into the 1940s, and the term "mongolism" was not finally replaced by "Down's syndrome" until the 1970s.[41]

Down's interpretation persisted, in the face of much evidence and expert opinion against it, because it fit into a common narrative for understanding both disability and evolutionary progress. Despite rejection by many experts of his specific racial typology, the notion that intellectual disability was a matter of the body's reverting to, or being arrested at, an earlier stage of evolution was entirely mainstream. Charles Darwin himself had mentioned in *The Descent of Man*, under the heading "Arrests of Development," that idiots "somewhat resemble the lower types of mankind."[42] The American geologist John Peter Lesley in 1868 used intellectual disability as evidence for the human evolution: just as some looked to Africa for examples of intermediate forms between ape and man, Lesley pointed to the "millions of idiots and cretins," products of "arrested development," as living "missing links."[43]

In general, educators and other professionals who worked with intellectually disabled people increasingly moved toward using terminology steeped in modern notions about time, progress, and competition. Older terms persisted, such as "feebleminded," sometimes as a catchall term and sometimes as a category placed just above "imbecile" and "idiot," but joining these at the turn of the century were terms implicitly or explicitly tied to notions of time: "backward," "atavistic," "mentally inefficient." The term that eventually became standard was "mentally retarded."

The verb "to retard" dates at least to the fifteenth century in the sense of "to slow down or hinder," and it was used in a variety of contexts in the nineteenth century, perhaps most often in the common phrase "to retard progress." In the 1890s the *Journal of Sociology* could publish an article titled "A Retarded Frontier" about the inhabitants of mountain communities in eastern Kentucky. The title phrase was never explained, because readers would not find it perplexing or out of place. The author wrote blandly of "the interesting field for social study which this retarded frontier affords" and hoped that social scientists would investigate "this curious social survival" before it "lost its comparatively primitive character . . . arrested at a relatively early stage of evolution.[44] No doubt the inhabitants of these communities would not have been pleased by the characterization, but the word did not yet carry the charge and particular connotations it would later acquire. In the twentieth century, however, "retarded" followed the same path as "handicapped," from general multipurpose adjective to a stand-alone noun: "the mentally retarded." Like "handicapped," its usage tracked the changing understanding of the significance of time as the problem shifted from a mind that was "feeble" to one that could not keep up with the competition.

Educators first began using "retarded," as shorthand for "grade-retarded" in the 1890s to describe children who did not advance with their age cohort, for reasons ranging from lack of discipline to malnutrition to mental defect.[45] If the broader context for this terminology was a competitive industrializing economy, the more proximate impetus was the creation of a compulsory, graded public school system. Graded public schools, which gradually replaced multiage grouping or "one-room schoolhouses" in the second half of the nineteenth century, were structured on a competitive model in which children were ranked in comparison with others of the same age. Leading students were honored, the mass in the middle were rewarded with advancement to the next grade, and those at the rear of the pack were demoted or removed entirely from the regular classroom. As Penny Richards has noted, children with intellectual disabilities typically had been included in ungraded classrooms, "allowing some children to progress more slowly without drawing unwelcome attention (or requiring special accommodation)." With the advent of graded schools, however, children who did not keep up became a problem.[46]

One of the functions of the graded school system was to instill the values and skills called for in a competitive economy: punctuality, the ability to adhere to a schedule, respect for authority, and an individualistic and internalized work ethic. Another was to sift and rank students on the basis of those values and skills. Quick thinking was rewarded by testing practices that made speed one of its basic elements. Feebleminded children were defined as "unable to compete successfully with normal children in their school work" or as those who "retard the work and interfere with discipline."[47] The ability to compete in school was used as a predictor of ability to compete in the world. When grown, feebleminded adults would likewise "not able to compete on an equality with normals."[48] Persons who had never been considered feebleminded became so because they fell behind in the race.[49] By 1909 the Bureau of the Census reported that part of the generally accepted definition of feeblemindedness was "arrested or imperfect mental development as a result of which the person so affected is incapable of competing on equal terms with his normal fellows."[50]

Educators for a time maintained a distinction between the "merely retarded" child and the "actual defective." For example, the superintendent of the Philadelphia public schools in 1908 distinguished between children with "serious physical or mental defects" and those who "are retarded on account of difficulty with the language" or whose "retardation is due to the fact that pupils are already over age on entering school." Retardation might also be attributed to the "inefficiency of a particular teacher, principal, superintendent, or administrative system."[51] Even as late as 1913, an education professor was within civil bounds when he cautioned that paternalistic leadership by school administrators "tends to produce retarded teachers and so retarded students."[52] Retardation spoke to progress in school, not identity.

While the term continued to be used in the older sense of "grade retarded" for several decades, at the same time a subtly different usage began to appear alongside it. Educators increasingly applied it not just to a child's progress, but to a quality that defined a particular category of persons—not the merely grade retarded, but the mentally retarded. One educator used "retardation" to refer to the "failure of many pupils to be promoted regularly from grade to grade" while in the same article referring also to "mental retardation," without making a definite distinction between them.[53] Others took pains to maintain two distinct meanings for "pedagogical retardation" and "mental retardation."[54] With increasing reliance on standardized testing, however, the term "retarded" gradually came to denote less an educational standing than an overall mental status. Children who were grade retarded when they could not keep up with their cohort became mentally retarded when they brought up the rear in a competitive testing system. Finally, they became retarded in the evolutionary race when heredity and eugenics entered the conversation.

The concept of mental retardation originated in the field of education but soon spread to others. In the journals and proceedings of the medical professionals who ran institutions for the feebleminded, the term appeared now and then up through the turn of the century, usually to describe various ways in which progress or growth might be impeded and only occasionally as "retarded mental development."[55] By the 1910s and 1920s, however, the term had become standardized in phrases such as "retarded intelligence," "the mentally retarded," and the "abnormally retarded," as an umbrella term for intellectual disability. Transformed into a description of a type of person, "retarded" filled the need for a label that captured the idea of someone who was both uncompetitive economically and a laggard in evolutionary development.

In 1923 Stanley Porteus, a professor of psychology at the University of Hawaii and a leading researcher on intellectual disability, advocated investigation into the parallels "between the mentally retarded individuals in our own race and the groups of mankind that are racially retarded." Porteus wanted to determine whether "the feebleminded represent survivals of inferior stock within the one race just as primitive peoples represent survivals of inferior racial stock." If so, the policy implications would be profound, for it would demonstrate that "feeblemindedness is so deeply rooted biologically speaking, that the only solution of the problem would be found in a eugenic program that would cut off the supply of feeblemindedness at its source." Conversely, if it could be determined that "primitive peoples are merely mentally retarded communities . . . then the task of attempting to raise them by educational and missionary work would also be a hopeless one."[56]

By the 1910s educators were increasingly employing "social efficiency" as a key concept and describing it as the preeminent goal of education. One of the leaders of the social efficiency movement, Leonard Ayres, published an "Index of Efficiency" in 1909. Designed to determine "the relation of the finished product to the raw material" in schools and also "conditions of maximal theoretical efficiency," it was a method whereby, according to Herbert Kliebard, "the production metaphor applied to the curriculum could be used with ruthless precision."[57] Educators and social scientists began to speak of mental retardation as a matter of "mental and social inefficiency" or "extreme inefficiency."[58] Exclusion from the general classroom became a question of "the point where the inefficiency of the individual becomes so great" that he could not keep up with "the intellectual activities of the group to which his age would naturally assign him." While intelligence testing might be incapable of "an exact statement of the efficiency of any individual," it could nevertheless "select out the most inefficient," allowing teachers to distinguish those whose families were "of normal mentality" and therefore might "grow into normal efficiency" from those "poor intellectually befogged creatures" descended from "a feeble-minded, inefficient clan." Tests at least provided a "roughly drawn line between inefficiency and extreme inefficiency." The ability to process information quickly was a crucial ingredient in the emerging definition of intelligence, for one of the "salient characteristics of the mental defective is never to do anything regularly and on time except through training and habit formation or from outside compulsion." Intelligence tests "brought out a striking difference in time of performance," revealed "striking differences in efficiency," and allowed standardized comparisons within age groups to "the grade of an efficient child."[59]

For many economists, the terms "defective" and "inefficient" were practically synonymous.[60] The Columbia University professor Henry Seager, among others, made this an explicit aspect of his argument for a minimum wage, arguing in 1913 that low wages resulted from too many inefficient workers competing for jobs with more able ones. Employers hired them because the low wages they demanded outbalanced their inefficiency, which in turn dragged down the wages of all workers. A minimum-wage law, by making them unemployable, would "extend the definition of defectives to embrace all individuals, who even after having received special training, remain incapable of adequate self-support. Such persons are already social dependents. The plan merely compels them to stand out clearly in their true character." In the long run, eugenic measures would be needed "to prevent that monstrous crime against future generations involved in permitting them to become the fathers and mothers of children who must suffer under the same handicap." If Americans wanted to "maintain a race that is to be made up of capable, efficient and independent individuals and family groups we must courageously cut off lines of heredity that have been proved to be undesirable."[61]

Moral defect was likewise discussed in these terms. The "loss to honest industry due to the reduced efficiency of sexual perverts," for example, was one of the costs of immorality identified by the national Moral Efficiency League.[62] Cities such as Pittsburgh established "morals efficiency commissions" to remove moral impediments to the smooth working of the city,[63] educators called for schools to promote social efficiency via courses in moral efficiency, and religious leaders called for greater moral efficiency via expanded religious education. Rabbi Emil G. Hirsch, for example, worried that although "efficiency is the keynote of modern man's ambition in all fields of human activity" and "the dominant passion" among educators, "progress toward greater moral efficiency has not kept pace." Indeed, "it has been retarded" by misplaced priorities.[64]

Specialists in fields related to disability often spoke of immigration policy as an issue they had a duty and the expertise to address.[65] A 1922 Michigan study, for example, investigated the "underlying cause" of higher rates of grade retardation among the children of immigrants. The researchers administered intelligence tests to determine whether these "differences in retardation might be the result of differences in intelligence." Unsurprisingly, they found that school success was correlated with the test scores, both of which were correlated with ethnicity, including the "interesting fact that all the Germanic groups—Norwegian, German, Swede, English, and Austrian—test higher than any of the non-Germanic groups." The study suggests how grade retardation came to be conflated with mental retardation. Success in school and intelligence test scores were treated as if they measured two distinct kinds of things, with intelligence the cause and school success the effect. Not considered was the possibility that both might be measuring the same types of skills, including the ability to work quickly. Scores were then mapped onto ethnicity, demonstrating that grade "retardation according to nationality follows very closely the median intelligence quotients of the nationalities." Extending the discussion to immigration policy, the author extrapolated that the "admission into this country of large numbers of immigrants of relatively low intelligence" meant that "in coming generations the general intelligence of the American people is likely to be lowered."[66]

Similar conversations took place among immigration officials. According to Howard Knox, an Ellis Island medical officer, the most numerous and dangerous of undesirable immigrants were the "great class of defectives" who often went through daily life unrecognized by laymen as defectives, "the morons and constitutional inferior types." Both had several attributes in common, including an "undeveloped moral tone" and an "inability to compete unaided in the struggle for existence." These undetected defectives would undoubtedly "start a line of defectives whose progeny, like the brook, will go on forever, branching off here in an imbecile and there in an epileptic."[67] Knox went on to develop a series of intelligence tests for use at Ellis Island and a *Manual for the Mental Examination of Aliens* for the Public Health Service, both of which became essential tools in exposing defective immigrants who were doing their best to pass as normal.[68]

Normal

A medical officer stationed at Ellis Island warned in 1906 that in cases of poor physique there was invariably "some abnormality in the individual's mental and moral make-up."[69] At a 1910 immigration hearing for Bernhard A. Mydland, who had curvature of the spine, a board member asked the examining physician, "How near normal is he?"[70] In 1908, in the case of Arotioun Caracachian, certified for "arrested sexual development," the examining physician testified that people with his condition generally "do not come up to the normal standing" mentally.[71]

The use of the word "normal" to describe the person ostensibly without defect dates from the late nineteenth century. Although not immediately apparent, in context it too was an evolutionary concept. Until early in the century, the word was most often used in reference to the "normal" or "right" angle. Georges Canguilhem has described how physicians adopted it early in the nineteenth century to describe an organ or organism, which was in its "normal state" when functioning properly and an "abnormal state" when not. (In a parody of medical discourse, Balzac wrote in *Eugénie Grandet* that Mademoiselle d'Aubrion had "a nose that was too long, thick at the end, sallow in its normal condition, but very red after a meal.") In the 1820s Auguste Comte imported the term into sociology to describe the healthy state of a society.[72] From there it migrated from one discourse to another. Political economists in the 1850s, for example, defined the "normal nation" as economically mature and capable of self-government, at times likening it to the independent and self-sufficient "normal man." The abnormal nation, needing "protective, restrictive, or reformatory" supervision, was likened to "the child or the idiot, the spendthrift or the sot."[73] The "normal schools" (derived from the French *école normale*) established in the United States beginning in the 1830s were so called because they set common standards or norms for public schools, and by 1862 an *Atlantic Monthly* writer could play with the unstable meanings of the word: "Your Normal schools wun't turn ye into Normals."[74]

By the end of the century, according to Ian Hacking, normality had "displaced the Enlightenment idea of human nature as a central organizing concept" in Western societies; rather than inquire into human nature, the modern question was what was normal for people to be and to do.[75] In earlier years, disability was often discussed in terms of what was natural and what was unnatural. Children born with physical anomalies might be described as "monsters," an unnatural product of sinful and unnatural practices. This usage was ubiquitous in metaphorical speech, much as normality would later become. For example, Edmund Burke, in *Reflections on the Revolution in*

France, drew a rhetorical contrast between the natural constitution of the body politic and the monstrous deformity that the Revolution had brought forth. Burke repeatedly referred to "public measures . . . deformed into monsters," "monstrous democratic assemblies," "this monster of a constitution," "unnatural and monstrous activity," and the like (not to mention his many references to the "madness," "imbecility," and "idiocy" of the revolutionary leaders). This rhetoric of monstrosity was not peculiar to the conservative cause. Tom Paine, in his response to Burke, also found the monster metaphor apt and useful, but he turned it around: "exterminate the monster aristocracy," he wrote.[76]

The natural and the normal both were ways of establishing the universal, unquestionable good and right. Both were constituted in large part by being set in opposition to culturally variable notions of disability—just as the natural was meaningful in relation to the monstrous and the deformed, so were the cultural meanings of the normal produced in tandem with defect.[77] The concept of the normal functioned simultaneously as both description and prescription. In its older usage, a normal angle for a builder was both a type of angle and the desirable angle; in all of its subsequent applications, from medicine and sociology to eugenics and education, the normal was also the desirable. Thus, the *New Orleans Delta* in 1856 could maintain that the people of the northern states would "never exist happily and normally" until they too adopted a system of labor based on slavery.[78] Francis Galton in 1877 could use it as a neutral description of a distribution that "is perfectly normal in shape," but also to stigmatize people who fell to one end of a distribution as "unfit" and on that basis advocate their sterilization.[79] While it ostensibly denoted merely the existing average, normality functioned as a standard that compelled conformity—"abnormal" typically signified only an undesirable deviation. To ask, "Is the child normal?" was rarely to express concern about *above*-average abilities. "Abnormal" signified the subnormal.[80]

The complex etiology of the modern use of "normality" included the rise of the social sciences, the science of statistics, and industrialization, with its need for interchangeable parts and interchangeable workers. Mass production wedded to mass consumption made standardization a pervasive aspect of modern life. Moreover, the power of the idea of normality arose in the context of a growing cultural belief in progress. The ideal of the natural had been a static concept for what was seen as an essentially unchanging world, dominant at a time when the "book of nature" was represented as the guidebook of God. The natural was good and right because it conformed to the intent or design of nature or the Creator of nature. Normality, on the other hand, was an empirical and dynamic concept for a changing and progressing world, based on a belief that one could discern in human behavior the central tendency and direction of human evolutionary progress and use that as a guide. The ascendance of normality signaled a shift in the locus of faith, from a God-centered to a human-centered world, from a culture that looked within to a core and backward to lost Edenic origins to one that looked outward to behavior and forward to a perfected future.

Socially, normality gained potency from the growing acceptance of theories of evolution that, as George Cotkin put it, "posited change, process, and struggle as essentials" and simultaneously equated evolution with progress.[81] In the context of an assumption that the normal tendency of the human race was to improve steadily, to advance ever further from its animal origins, normality was implicitly defined as progressive. Normality on the individual and social level contributed to forward motion, while abnormality was a retarding or atavistic force with the potential to slow or even reverse progress. By the late nineteenth century, both nonwhite "lower" races and defective individuals were similarly described as evolutionary laggards or throwbacks. A popular 1873 geography textbook, under the heading "The White Race the Normal, or Typical, Race," compared the "harmony in all the proportions of the figure" of the white race with the degenerate races. African Americans, moreover, were said to flourish in their "normal condition" of slavery, while the "free or abnormal negro" succumbed to illness, degeneration, and inevitable extinction.[82]

James W. Trent has argued that the 1904 World's Fair displays of "defectives" alongside displays of "primitives" evinced related schemes of classification. Defective individuals and defective races were both ranked on the basis of how "improvable" they were—capable of being educated on the one hand or civilized on the other—and both were explained in terms of atavism or lack of evolutionary development.[83] Teachers of the deaf at the turn of the century spoke of making deaf children more like "normal" people and less like "savages" by forbidding them the use of sign language, which "resembles the languages of the North American Indian and the Hottentot of South Africa."[84] They argued that when "man emerged from savagery he discarded gestures for the expression of his ideas,"[85] and that ever since spoken language had been "the normal and universal method of communication," with the exception only of "tribes low in the scale of development."[86] To use sign language with deaf children would "push them back in the world's history to the infancy of our race."[87] Given that "speech is better for hearing people than barbaric signs, it is better for the deaf; being the 'fittest, it has survived.'"[88] One educator described a young girl who "had just broken out in speech" and experienced an "elevation in the scale of being."[89] Another argued that mute children could "gradually rise" if taught to speak, even if their development remained "retarded."[90]

Edward Seguin, the leading expert on intellectual disability in the nineteenth century, used the term "normal" in the medical sense during the 1850s and 1860s: "Sometimes the brain of idiots presents no deviation in form, color, and density from the normal standard; it is, in fact, perfectly normal."[91] By the 1880s, however, his colleagues were using it to describe the totality of a person, as when Isaac Kerlin, superintendent of the Pennsylvania Institution for Feebleminded Children, wrote that education of the feebleminded should effect "conformity to the habits and actions of normal people."[92] Teachers of deaf students discussed how their work differed from "ordinary work with the normal child," how they could shape "the deaf child into as nearly a normal individual as possible," and how their students could learn to speak in "the manner in which the normal child acquires speech."[93] The normal child could keep up successfully in the school system, the normal adult could compete as an independent individual in the economic struggle for existence, the normal nation was industrialized and competitive in the global market, and the normal race was competitive in the evolutionary race.

Normality was always a relative term, one that had meaning only in the context of competition within a group. The prominent psychiatrist Aaron J. Rosanoff told the Eugenics Research Association in 1914, "Who is to be counted as normal, who is abnormal, is the great question." The answer depended on comparison with others in a given society: "There can be no doubt that many persons, say, in Arkansas, New Mexico, or in Oklahoma, who are at large, and whom[,] moreover, none of their fellow citizens would consider as proper subjects for an insane asylum, would be promptly committed if they took up residence in Massachusetts or New York."[94] But at that same meeting Howard Knox, recounting his encounter with an immigrant "closely resembling the 'Missing Link,'" relied on the more common view of a normality defined in relation to human evolutionary competition more generally. "One familiar with the reconstruction of the man of the stone age could not help but note the close resemblance," he noted, describing the relevant details of the man's head and body. The immigrant had worked as a linesman for a telegraph company in his native Finland, an occupation "particularly well chosen for his physical make up, since he may have inherited the characteristics of his ancestors who perhaps often found it necessary to climb to the treetops to escape some giant animal of their time." The man was refused entry, Knox explained, because he was "one of a great class who possess atavistic features, indicative of a physically retrogressive make-up that is not good for our racial type."[95]

Defect and Degeneracy

Two closely related terms that also came into common use at this time, "defective" and "degenerate," like "handicapped," "retarded," and "abnormal," had long been applied in other contexts before acquiring meanings specific to disability. The more general term of the two, "defect," had long been used to refer to any kind of imperfection or lack. Shakespeare, for example, covered most of the possible defects to which humans are susceptible in *Venus and Adonis*:

> Were I hard-favour'd, foul, or wrinkled-old,
> Ill-nurtur'd, crooked, churlish, harsh in voice,
> O'erworn, despised, rheumatic, and cold,
> Thick-sighted, barren, lean, and lacking juice,
>
> Then mightst thou pause, for then I were not for thee;
> But having no defects, why dost abhor me?[96]

In the Civil War song "The Invalid Corps," an army physician rejects an enlistee, saying:

> You're not the man for me,
> Your lungs are much affected,
> And likewise both your eyes are cock'd,
> And otherwise defected.[97]

Mark Twain, in an apparently original coinage that never caught on, wrote in 1883 of a man who was "endowed with an extraordinary intellect and an absolutely defectless memory."[98] By 1899, however, when a *Popular Science Monthly* writer argued that "laws preventing the marriage of defectives and of their immediate descendants would go far to stem the tide of harmful heredity," readers knew that these "defectives" were not merely those who might be "lacking juice," forgetful, or cockeyed.[99] Rather, they were understood to pose social and eugenic threats to progress.

The sense of constant, inexorable change in modern times held both promise and danger. Americans at the turn of the twentieth century often expressed the hope that their nation was destined to lead the world on an upward path of unending progress, but just as often voiced apprehension that the nation faced imminent decline. The economy was expanding (albeit with a distressing boom-and-bust cycle), and a stream of new inventions were transforming everyday life, but all the while public figures warned that where the superficial observer might see progress, the more discerning saw deeper currents threatening to drag the nation, and perhaps the human race, back to an earlier stage of evolution. If humanity had risen from the apes, it could just as easily decline. The fear was not that decline might take humanity all the way back to its simian roots, but rather that one segment of humanity—"the civilized races"—might weaken and degenerate back to some intermediate, less civilized stage, or be overwhelmed like Rome by barbarian hordes.[100]

"Degeneracy" encompassed the concept of defect but had the added connotation of hereditary decline, such that a mild defect might within a few generations become a thoroughgoing "constitutional psychopathic inferiority."[101] Crime, pauperism, sexual immorality, violence, and indeed most of the social problems of the day were feared to be its rapidly growing consequences. A medical officer stationed at Ellis Island cautioned his colleagues that immigrants who had "poor physique" invariably also showed "signs of physical degeneracy."[102] Restriction advocates were especially anxious about the dangers. Thomas Darlington, president of the New York City Board of Health, warned in 1906 that because of the flood of immigrants of poor physique, "we are in danger of becoming a degenerate nation."[103] A New York surgeon and medical researcher, Arthur L. Fisk, wrote in 1911 that an "ever increasing horde of degenerates from all nations is entering our fair land, who are debasing the physical, mental and moral being of the nation."[104] The admission of degenerate immigrants was, as Robert DeCourcy Ward put it plainly, "a crime against the future."[105] The Yale economist Irving Fisher wondered whether the American nation might not already be in general decline; the "common opinion is undoubtedly that we have made great progress and are making great progress now," but this very "same opinion was held, so historians tell us, just before the down-fall of Rome and of other civilizations which have failed." He feared that "much of what we call progress is an illusion and that really we are slipping backwards while we seem to be moving forwards." The United States, where the labor-hungry economy gave "undesirable citizens far greater opportunity to multiply than they had at home," faced the most immediate risk, but "injuring the germ plasm of the human race" would eventually result in "populating the earth with defectives, delinquents and dependents."[106]

Another sign of eugenic crisis was the decline in the birthrate among the educated classes. The term "race suicide," coined by the sociologist Edward Alsworth Ross in 1901 and popularized by Theodore Roosevelt, expressed the fear that the best were being outbred by their inferiors. Superior races, by virtue of their greater prudence and foresight, produced fewer offspring, Ross believed, and thereby the "higher race quietly and unmurmuringly eliminates itself."[107] Roosevelt saw the same problem but cast prudent family planning in an even less favorable light; people who prefer few or no children, he told the National Conference of Mothers in 1905, "whether from viciousness, coldness, shallow-heartedness, self-indulgence, or mere failure to appreciate aright the difference between the all-important and the unimportant," deserved "contempt as hearty as any visited upon the soldier who runs away in battle."[108] Eugenicists such as Davenport warned that not only were "defective conditions" transmitted in the "germ plasm"; worse yet, they were "being reproduced faster than the more normal characteristics." Davenport attributed the difference in birthrates not to greater prudence among the better sorts, as did Ross, nor to their moral failings, as did Roosevelt, but to the inherent "erotic characteristics of the degenerate classes," which resulted in "a frightful fecundity" that far surpassed that of their superiors.[109]

Another worrisome aspect of degeneracy was its role in the causation of crime. American criminologists, following the pioneering Italian Cesare Lombroso, theorized criminality as a reversion to an earlier stage of evolution, evident in a primitive sense of morality as well as apelike physical characteristics such as a sloping forehead, facial asymmetry, disproportionately long arms, and insensitivity to pain. They worried that the present immigration included a great number of the "criminal type," knowing that "atavistic tendencies are more necessarily pronounced in the offscourings of older social systems" such as those in Asia and southern and eastern Europe.[110] The eugenics researcher and psychologist Henry Goddard argued that criminality was the result not of reversion per se but rather of arrested development; the morality of those he termed "morons" had failed to advance beyond a primitive stage of development owing to some hereditary defect. He believed such people to be especially dangerous because they often passed as normal—the "public is entirely ignorant of this particular group," he warned, and the "public school systems are full of them." Although the schools vainly attempted "to make normal people of them," inevitably their "primitive" instincts led them into a life of crime. Whether caused by arrested development or reversion, crime and immorality were failures to keep up in the evolutionary race.[111]

"Moral imbecility" was especially associated with this danger. Martin Barr, chief physician at the Pennsylvania Training School for Feebleminded Children, described it as "the most insidious and the most aggressive of degenerative forces; attacking alike the physical, mental and moral nature," the hereditary taint was everlasting—"lying latent it may be, but sure to reappear."[112] For this reason, the Ellis Island commissioner William Williams insisted that the bureau had "no more important work to perform than that of picking out all mentally defective immigrants." He dismissed the concern that they "may become a burden on the taxpayer" as a "relatively unimportant consideration." For Williams what was "vitally important" was that "such persons contribute largely to the criminal classes and that they may leave feebleminded descendants and so start vicious strains leading to misery and loss in future generations."[113]

Howard Knox worried also about those on the opposite tail of the bell curve, warning that too often "insanity and genius go hand in hand." He advised immigration officials to be on the lookout not only for the feebleminded but also for the "the high-strung, queer, eccentric person, or, as we medical men call them, 'the psychopathically predisposed.'" Although "intellectually they may be very bright and have [the] ability to solve mathematical problems and pass all the complicated tests that can be given them," they were nevertheless unsound, for it was "rather in the emotional sense that they are defective." Their particular type of defect made them "impulsive, thoughtless and given to the gratification of the baser passions, resulting finally perhaps in murder and other crimes." He believed that this "type is really to be feared more than the moron, or simple feebleminded person of more stable emotions," and it was therefore "the aim of those who examine immigrants to detect just this class."[114]

The idea of the evolutionary throwback, whose primitive morality was out of place and time in the modern world, was not confined to journals of criminology but also appeared often in popular literature. Jack London described a character in *The Sea-Wolf* as "a magnificent atavism, a man so purely primitive that he was of the type that came into the world before the development of the moral nature. He was not immoral, but merely unmoral."[115] Edgar Rice Burroughs used degeneracy as a simple tool of character development, for example in the case of "degenerate Charlie," a.k.a. "Dopey Charlie," the criminal tramp.[116] It was part of Sherlock Holmes's tool kit for recognizing the criminal type.[117] Bram Stoker's *Dracula* reflected the "wider debate on the physiognomy of the 'born criminal,'" according to the historian Daniel Pick, as well as "the moral panic about the reproduction of degeneration, the poisoning of good bodies and races by bad blood, the vitiation of healthy procreation."[118]

The notion that almost any trait might be heritable extended to the fear that democracy itself was imperiled by immigrants who had no hereditary capacity for it. In some cases the defect was characteristic of entire races. Francis Walker wrote in 1896 of his "grave apprehension" about the great number of immigrants who possessed "no inherited instincts of self-government and respect for law," lacked the necessary "aptitudes which fit men to take up readily and easily the problem of self-care and self-government," and consequently could only be governed by "the club of the policeman or the bayonet of the soldier." They were "beaten men from beaten races; representing the worst failures in the struggle for existence."[119] Asians in particular were said to lack a hereditary capacity for democratic governance. When discussing European immigration, however, it was usually some form of individual degeneracy that preoccupied restrictionists. Davenport, for example, believed that the anarchist suffered from an inherited pathology such that "his brain may be incapable of developing properly," leaving him with "defective judgment, memory and, even, instincts, unable to appreciate the traditions of human society or, perhaps, impelled constantly to run counter to the fundamental principles of that society—tearing them to shreds." Such an individual additionally "may suffer from melancholia or paranoia in its multifarious forms, leading him to commit arson or murder and to assassinate high officials. Heavy is the toll human society pays for the presence of these degenerates."[120] The Public Health Service physician Allan McLaughlin thought that Americans feared "not only the germ of bodily disease, but the germ of anarchy and also favorable media for its growth," in the form of immigrants mentally susceptible to radicalism.[121] Gustavus Doren, superintendent of the Ohio Institution for Feeble-Minded Youth, worried too that "unless preventive measures against the continuously progressive increase of the defective classes are adopted, such a calamity as the gradual eclipse, slow decay and final disintegration of our present form of society and government is not only possible, but probable."[122]

The Contagion of Genes

Historians of immigrant medical inspection have focused mainly on the question of communicable diseases.[123] The iconic images of inspection are those depicting the dreaded eye inspection, in which the eyelid was pulled up and out, at first with fingers and later with button hooks dipped in bleach, to reveal telltale signs of trachoma. The danger from contagious disease was only part of the motivation for the inspection process, however. Contagion of genes was seen as at least an equal danger and, after the turn of the century, as by far the greater threat.[124] The journal *Medical Record* editorialized in 1912 that screening for mental defects was "of vastly greater concern to this country than the quarantining of cholera, smallpox, plague, or any other infectious disease," for while epidemics always come to an end, "the admission of mentally defective immigrants strikes at the very roots of the nation's existence . . . leading to incalculable misery, pauperism, and criminality in future generations."[125] The Public Health physician Alfred C. Reed made explicit the parallel between the contagion of germs and the contagion of genes in 1913 in the course of a discussion of immigration restriction and public health: "Disease or defectiveness of mind or body in the immigrant must be considered from two standpoints. . . . First is the immediate result on those with whom the immigrant comes in contact. Second is the effect on the descendants of the immigrant, and indirectly on the general public; in short, the eugenic aspect."[126]

Martin Pernick, writing on the convergence of eugenics with public health during this period, has described the ways in which heredity and contagion were spoken about in similar terms. Infections, for example, were carried by germs and inheritance by germ plasm, and "in both cases, 'germs' meant microscopic seeds" that could "spread contamination from the bodies of the diseased to the healthy." Furthermore, the common idea of "blood as a medium of infection" was easily conflated with blood as an "age-old metaphor for heredity." In both instances, "'bad blood' meant you were contaminated and contaminating, whether the specific agent was a germ or the germ plasm." If the rhetoric of causation was remarkably similar, so was the rhetoric of cure. Eugenicists called for the segregation of defectives in institutions to prevent the "spread of hereditary disease," a practice that "directly echoed the centuries-old effort to stop the spread of infections through quarantine." The word "segregation" itself "first was used medically in the mid-nineteenth century to mean 'selective isolation' or 'quarantine.'" Finally, both germs and germ plasm could be "stopped from spreading by a new method called sterilization," used by both eugenicists and bacteriologists "to eliminate the agents that reproduced disease."[127]

Hereditary defect and degeneracy were, moreover, a problem not only of germ plasm handed down from the past, but of the ongoing creation of new defective inheritance through the passing along of acquired characteristics. Biologists had been moving toward an understanding of genes as insulated from and unaffected by environmental conditions or the behavior of the individual ever since August Weismann showed in 1889 that mice with their tails cut off did not bear short-tailed offspring. Many American biologists, however, not to mention the public, legislators, and government officials, continued to adhere to neo-Lamarckian ideas. It was not until the "great synthesis" in the 1930s of Darwinian natural selection with Mendelian genetics that the inheritance of acquired characteristics finally disappeared from scientific discussion. The possibility that acquired characteristics could be inherited made discussion of the dangers of unrestricted immigration far more expansive than it might have been otherwise: if acquired characteristics could be inherited, then any defect, no matter its origin, was potentially dangerous to the public health.

When immigration officials confronted immigrants they thought to be degenerate, the history of the immigrant and the future of the nation were uppermost in their minds. In 1906 Emil Letz, a fourteen-year-old German immigrant, was excluded not merely for his criminal acts since arriving in the country six months earlier, but for being a "hereditary degenerate." Letz's uncle had paid for his passage from Europe and taken him into his home in St. Louis but was now eager to send him back. Letz had twice stolen money from him, in partnership with the uncle's eighteen-year-old son, and the two had run away together. When Letz returned, he was arrested and committed by a juvenile court to a reformatory, the St. Louis Industrial School. Letz admitted to the immigration officials that the charges were true but claimed in his defense that he had only been the lookout—that his cousin had done the stealing and kept most of the money. "I have always been easily influenced and anybody that talked with me could coax me to do anything," he lamented. The case against him briefly described his actions but devoted many more detailed pages to his defective heredity, as testified to by the estranged uncle: the boy's grandfather was a "confirmed drunkard," his father "a cripple, having a withered arm. . . . The arm is now shrunken to mere skin and bone, the fingers are about as large as pencils." His mother had "always been sickly" and was "also afflicted with some nervous disorder and is considered weak minded." Two uncles had in the past "had trouble with the police." The uncle testified that he was convinced that Letz was "a victim of natural criminal tendencies" and that he had seen "abundant signs in his nervous and moody disposition to believe that he is a degenerate or weak minded." The Public Health Service physician agreed and certified him a "hereditary degenerate of criminal tendencies." In addition to the "family history of pauperism, mental and physical disability, and degeneracy," the report noted that Letz was "undersized," and that "the causes of said alien's degeneracy and present condition have existed long prior to his landing," namely "moral and physical deficiency and degeneracy of parents and other members of family."[128] Letz had committed crimes, but the bulk of the evidence seen as relevant against him concerned his defective heredity.

For many Americans, heredity was the key to an array of social problems. No one knew the limits of what was heritable, and given that uncertainty, even the most alarmist predictions were accorded serious consideration. Although "positive eugenics" was one element in the public discourse—that is, the importance of encouraging individuals with superior qualities to reproduce—"negative eugenics" was what captured the imagination most vividly. Popular fears mounted that immigrants were introducing defects into the national body that would spread through it like an infection, bringing in its train ever-increasing crime, pauperism, and other social pathologies. What was heritable? Davenport wrote in 1909 that "we know already that *many* human characteristics are inherited in mendelian [*sic*] fashion—polydactylism, syndactylism, short fingeredness, bleeding or haemophilia, night blindness, congenital cataract, color blindness, keratosis palmse [*sic*], albinism, eye color, color and curliness of the hair." On that basis, he deduced that "doubtless many, if not all, of the elementary, physical, intellectual and moral characters are thus inherited." Laziness was one quality that Davenport had concluded from his studies to be almost surely inherited, the result of which was invariably pauperism. In his often inimitable prose, Davenport wrote that "it will be observed that we have not here to do merely with a high percentage of pauperism in the offspring of two lazy people, but with 100 per cent, or complete, pauperism. The children cannot rise in any particular quality above the potentiality of their more advanced parent."[129] An 1896 study reported that not only was "feeblemindedness" heritable, but that other "abnormalities, such as blindness, deaf-mutism, etc., were found to be several times as frequent" in families in which feeblemindedness was found. The authors warned that "by permitting the feeble-minded to reproduce, we are not only increasing their number, . . . we are also increasing the number of abnormals."[130]

Chaje Sara Schneiderofsky, a twenty-two-year-old Russian Jew, fell victim to this growing tendency to view individual characteristics through the lens of heredity. She had come to the United States in August 1905. In February 1908, by which time she had changed her name to Sarah Taylor, a warrant was issued for her arrest. After being admitted to the Philadelphia Almshouse Hospital, where she was diagnosed as suffering from insanity, she became a public charge. Because this had occurred within three years of her arrival, she could still be deported if it were shown that her condition was due to causes existing prior to landing. The physician at the hospital concluded that her condition was "due to heredity" and that "the patient was so far advanced in the disease" she must have been "mentally ill for some years." A Public Health Service physician examined her at the hospital and concurred. In these cases, he wrote, "a detailed family history is difficult to obtain," but "in this type of insanity defective heredity is conceded to be the chief underlying factor in about seventy per cent of the cases; and since defective heredity has its origin *ab ovo*, therefore the causes producing her present mental condition must have long antedated her arrival in this country." He did not explain the reasoning that allowed him to slide effortlessly from "seventy per cent" to "must" in the space of one sentence.

An immigration official went to the hospital to serve the warrant but "found this alien in a delirium of lunacy." Instead, he located her parents, whom he interviewed along with the physician who had treated Taylor. The inspector questioned the parents closely about whether anyone in their family history had suffered from "insanity or nervous diseases" or been committed to "an insane institution." There was no insanity in the family, the father testified; furthermore, the girl herself was "never sick" and had worked as "a finisher on men's clothes" until three months ago, when she fell off a streetcar. Soon thereafter, "one day she began to talk from heat [*sic*]; she did not do any damage, but talked to herself, whereupon she was sent to the hospital by the advice of the physician." The inspector asked, "Can you give any reason why your daughter should not be deported?" The father replied, "Because she has no one to go to over there, and here, when she will get better, I will take her back to my house and care for her." The mother added, plaintively, "How can we send her back when she has no one to go to?" The family's physician next testified that, following her accident, he had treated her for various physical injuries as well as "pains in the head, delayed menstruation, and insomnia and nervousness." Asked whether he had "noticed anything peculiar in her mentality," he said no, not at first, but more recently she had become "apprehensive," "did not comprehend questions properly or quickly," had "marked tremors of the fingers," and her "knee jerks were greatly exaggerated." When he last saw her, at the hospital, "she constantly complained of severe pains in the head."

The inspector's conclusion: "From the facts in this case, and the testimony submitted, it is evident that said Chaje Sara Schneiderofsky, alias Sarah Taylor, has become a public charge in this City, due to causes which existed at the time of her landing in the United States, and is, therefore, in this country in violation of law, and should be deported." The evidence he cited for his conclusion was the diagnosis of the hospital and Public Health Service physicians that she was "suffering from insanity, due to heredity." He added, however, that the father had filed a claim with the streetcar company for compensation for her injuries and recommended that deportation "be stayed pending the adjustment of her case; because to do otherwise, would entail hardship upon these very poor people."

Commissioner-General Sargent summarized the case for the Secretary of Commerce and Labor. There was the testimony of Taylor's personal physician that "the alien was of sound mentality previous to the accident on the street car," which would mean that she could not be deported under the law. "But there are two very satisfactory medical certificates in the case showing clearly that the causes of the alien's becoming a public charge existed prior to landing." Again, how the supposed 70 percent probability showed "clearly" that her condition was hereditary was left unexplained. It may be that the dread of hereditary defect was powerful enough to overcome that 30 percent of uncertainty, at least within the bureau. Fortunately for Taylor, it was not so for the Acting Secretary of the Department of Commerce and Labor, Charles Earl, who pointed out that the "conclusion of the medical officers is in conflict with the testimony" of the physician "who had the girl under observation several weeks," that "she has been here over two years and during that time showed no signs of insanity," and that the sequence of events indicates that "her present condition is the direct result of her injuries." He ordered the warrant for her arrest to be canceled.[131]

At the turn of the twentieth century, the growing tendency to see life as a race and progress as the result of an unending struggle for existence came to dominate how past, present, and future were understood. As analogies of competition became ubiquitous in every scholarly and professional discipline as well as the wider culture, the cultural understanding of disability was inevitably redefined in terms of this new and unsettling vision of time. Defectives were handicapped as individuals competing for employment, they were impediments to the nation in its international competition for economic dominance, and they were dead weights in the great quest for a superior American race. The critical task for immigration policy was to winnow out likely losers in the labor market and threats to the progress of the nation. The intense fear of human defect that characterized the immigration restriction movement grew, in good part, from the perceived demands of "these pushful days."

3

Dependent

Moische Fischmann arrived at Ellis Island in December of 1913. Thirty years old, Fischmann had supported himself as a blacksmith in his native Russia since he was fourteen. Now, in the midst of ongoing pogroms against Jews in Russia, he had decided to come to America to join his brother and sister in New York. The Public Health Service physician on duty, however, found Fischmann to be afflicted with a physical defect and sent him before the Board of Special Inquiry to determine his eligibility to land.

At his hearing, Fischmann was unable to testify. While the Bureau of Immigration employed a large staff of interpreters for the multitude of spoken languages used by immigrants, interpreters fluent in the sign languages of the world were not among them. Instead, Fischmann's brother and sister were asked to speak on his behalf. His brother told the board that he had lived in the United States for nine years and was now a citizen. He had a good job as an ironworker and savings of $235, a respectable sum for a working man at the time. His sister testified that she had been in the United States for six years, worked as a paper box maker, and had managed to save $455. Three cousins accompanied them, one with a letter from his employer guaranteeing Fischmann a job. All promised to support him through any difficulties while expressing confidence that, as a skilled blacksmith, Fischmann would have no difficulty becoming self-supporting. The board concluded, however, that Fischmann's "certified condition is such that he would have considerable difficulty in acquiring or retaining employment" and voted unanimously to return him to Russia.

Fischmann appealed. In addition to letters promising financial support from his relatives, Fischmann had an attorney provided by the Hebrew Immigrant Sheltering and Aid Society who argued forcefully that Fischmann was entirely capable of self-support. Moreover, he stressed that if Fischmann were excluded, he would be "sent back to misery and possibly destruction," reminding the bureau of the perilous situation for Russian Jews in 1913. A representative of the aid society sent a letter arguing that "the work of a blacksmith does not require the faculty of speech, but rather the possession of muscle and strength, and this he has." It concluded, "We do not see how he can be termed a person likely to become a public charge with all his relatives eager to help him . . . nor how the fact that he is a deaf-mute will affect his ability to earn a living, especially since he has earned his own support in the past." A second employer, the Fagan Iron Works, wrote to say, "I need a blacksmith. Send him over as soon as possible and I will start him at twelve dollars a week."

The family's congressman, a member of the Committee on Immigration and Naturalization, wrote to the commissioner general that it would be "exceedingly distressing if this alien were sent back to a land in which he has no relative and really no near friend, and kept from joining his nearest relatives here, who will indemnify the Government by bond or otherwise against the alien becoming a public charge." He stressed that, "by allowing him to land, the family simply will be united, while to deport him means to send this unfortunate alien adrift." The commissioner general, however, was not swayed. He responded to the congressman that "there can be little doubt that the applicant's certified condition will seriously interfere with his earning capacity" and in a brief letter to the Secretary of Commerce and Labor noted simply, "I have previously commented upon the undesirability of admitting deaf mutes." The Secretary concurred and ordered him deported.[1]

Deaf people were among the thousands of disabled immigrants turned back at American ports as defective and undesirable. They were excluded in large part because they were thought likely to be bearers of a defective heredity; this justification was discussed openly by advocates of immigration restriction, although it only occasionally emerged from the background in official immigration documents. The official justification for their exclusion was that deaf people were social dependents rather than contributors. The presumption of their dependency began in childhood. Whereas public education for hearing children was considered a right as well as a sensible civic and economic investment, for deaf children it was an altruistic act that came under the purview of state boards of charities. Even in adulthood deaf people were often regarded as perpetual dependents. At the hearings for deaf immigrants, responsibility for their financial support was generally assumed to rest not with them but with their families or other guardians, regardless of their age, education, or employment history.[2]

The presumption of dependence was never put to an empirical test by the Immigration Bureau, nor was any evidence ever adduced to show that deaf or otherwise disabled people were more likely than others to be unemployed or dependent upon public aid. In the most comprehensive review of census and other data, the sociologist Harry Best determined in 1914 that "the fact that the deaf are usually found capable of taking care of themselves should not be, after all, a matter either of doubt or wonder. They are for the most part, as we have indicated, quite 'able-bodied,' and but for their want of hearing are perfectly normal in respect to 'doing a job.'" He concluded that they were "not a burden on the community," that they were "wage earners in a degree that compares well with the general population," and that they "stand in need of little distinctive economic treatment from society."[3]

The same would not hold true for every disability or disabled individual, of course, as various disabilities interact in various ways with particular settings, occupations, and individuals. Blind people were (and are) less likely to be employed than the sighted—taken as a whole, that is.[4] Immigration officials, however, not only failed to seek evidence about the average employability of people with disabilities, they usually sought no such evidence in particular cases. In 1897, for example, the attorney for Cerillo Lesbinto questioned a medical officer's conclusion that Lesbinto was "unable to earn a living at ordinary work" due to being blind in one eye and having "defective vision" in the other: "Have you given him any test that would lead you to believe that he is unable to do that work?" The officer responded with exasperation, "How can I put a man to a test as to whether he is able to go out and get employment and make a living[?]" The attorney pressed the point: "Don't you understand that it is merely upon your examination that it is said that he could not perform ordinary manual labor?" Finally, members of the board intervened to protest that it was merely "common sense"—after all, one asked, what employer "would give him preference over able-bodied men?"[5] Officials were ostensibly assessing an individual's capacity for work, but really they were sorting individuals into a rough-and-ready typology, built on untested assumptions that were themselves justified mainly by the fact that they were widely shared.

Women

Common sense is not a reliable guide to the likelihood of, or the reasons for, unemployment among people with particular disabilities at particular moments in history.[6] Women at the turn of the twentieth century were also defined as dependents, despite the ubiquitous presence of working women. For both disabled and female immigrants, a kind of Platonic ideal of dependence prevailed, against which flesh-and-blood human beings were comparatively insubstantial and carried little weight. When Congress began to legislate on immigration policy, it was common sense and therefore not necessary to specify that single women should be more closely scrutinized than men when they attempted to immigrate unaccompanied. The public-charge provision excluded unaccompanied dependents, and women clearly and unambiguously came under its purview. Thus, even though no part of the law explicitly addressed single women, they were frequently excluded as likely public charges, far more often than men and often solely on the basis of their sex.[7]

The presumption of female dependence left unspoken in immigration law has been examined by historians, but not the extent to which it was rooted in the concept of disability.[8] In debates over women's proper place and role, the concept of disability was pervasive. One of the central rhetorical tactics of opponents of women's equality, for example, was to point to the physical, intellectual, and psychological flaws of women, their frailty, irrationality, and emotional excesses. Women either had disabilities that precluded participation in the public sphere, or they became disabled when exposed to the rigors that characterized men's lives. Edward H. Clarke, a Harvard professor of medicine, achieved fame for his 1873 book *Sex in Education; or, A Fair Chance for Girls*, in which he warned that women who were educated like men risked disease and disability. Overuse of the brain, he maintained, was responsible for the "numberless pale, weak, neuralgic, dyspeptic, hysterical, menorraghic, dysmenorrhoeic girls and women" of America, while an appropriate education designed for frail constitutions would ensure "a future secure from . . . derangements of the nervous system."[9]

Clarke's thesis was reiterated and reinforced by physicians for decades to come. The chair of the American Medical Association section on obstetrics and diseases of women, William Warren Potter, issued a warning at a meeting of the Medical Society of New York in 1891 that inappropriate education was incapacitating the American girl for motherhood: "her reproductive organs are dwarfed, deformed, weakened, and diseased, by artificial causes imposed upon her during their development."[10] Arthur Lapthorn Smith, a professor of surgical diseases of women at Bishop's University in Montreal, took to *Popular Science Monthly* in 1905 to warn that the "majority of women of the middle and upper classes are . . . physically disabled from performing physiological functions in a normal manner." In his experience, "most of the generally admitted poor health of women is due to over education," for it "takes every drop of blood away to the brain from the growing organs of generation" and "develops their nervous system at the expense of all their other systems."

If the educated woman did manage to conceive, her overtaxed brain came into competition for sustenance with the growing fetus, "and in this rivalry between the offspring and the intellect how often has not the family physician seen the brain lose in the struggle." The "mother's reason totters and falls, in some cases to such an extent as to require her removal to an insane asylum." Ultimately, women's aspirations to equality would lead to race suicide, for "nature punishes the man who has all the natural instinct cultivated out of him, just as it does the woman, namely, by the extinction of his race."[11]

Physicians adapted the same argument to the antisuffrage cause. When Almroth E. Wright, a British pioneer in bacteriology, published his *Unexpurgated Case against Woman Suffrage* in 1913, the *New York Times* announced the book's publication with some fanfare and extensive excerpts. On the subject of women's equality, there was "very little real difference of opinion among men," Wright claimed, but unfortunately, although "no man can close his eyes to these things . . . he does not feel at liberty to speak of them." The chapter titles indicate the direction of the argument: "Woman's Disability in the Matter of Physical Force," "Woman's Disability in the Matter of Intellect," "Woman's Disability in the Matter of Public Morality." He began with the familiar argument that since a government's authority ultimately depended upon "physical force alone," equal-rights claims must rest on a physical capacity that women lacked. Next, he maintained that the "woman voter would be pernicious . . . by reason of her intellectual defects." Finally, the emancipated woman was dangerous "to the State also by virtue of her defective moral equipment"—defective because incapable of extension beyond the home and family circle. As a result, "woman is almost without a moral sense in the matter of executing a public trust such as voting." The "imprint woman's sexual system leaves upon her physical frame, character, and intellect" necessitated her exclusion from the political sphere. Knowing this, "the doctor lets his eyes rest upon the militant suffragist" and "cannot shut them to the fact that there is mixed up with the woman's movement much mental disorder; and he cannot conceal from himself the physiological emergencies which lie behind."[12] The *Saturday Review* called Wright's book "the soundest and wisest discussion of woman suffrage yet written," although it really only reflected "what the majority of men and women today are thinking about the feminist franchise agitation."[13]

That same year, Charles L. Dana wrote a long letter to the *New York Times* to offer his professional opinion, as a leading neurophysiologist and professor of nervous and mental diseases at Cornell Medical College, that "woman suffrage would throw into the electorate a mass of voters of delicate nervous stability." The demands for woman suffrage were coming mainly from "aggressive" feminists who "are definitely defective mentally." In fact, "measured by fair rules of intelligence testing, I should say that the average zealot in the cause has about the mental age of eleven." If intellectual disability was the cause, psychiatric disability would be the effect: were women to "achieve the feministic ideal and live as men do, they would incur the risk of 25 per cent more insanity than they have now."[14] Antisuffrage activists such as Grace Duffield Goodwin borrowed from the physicians' vocabulary and argued that women's "temperamental disabilities," meager "endurance in things mental," and deficient "nervous stability" would lead to debility under the stress of public life; already, the "suffragists who dismay England are nerve-sick women."[15] The woman question, as it was called, was ultimately a medical problem that called for separate and special care, and those with the greatest authority on the subject were physicians. Their prescription? Particular care and special education adapted to women's special needs. The proper diagnosis and remedy for the "problem of woman's sphere," wrote Clarke, "must be obtained from physiology, not from ethics or metaphysics."[16]

Advocates of equality shared the same premise that disability justified inequality, reflecting a widespread consensus. They argued that women were not disabled and therefore deserved equality, or in a few instances that women were not inherently so but had been made disabled by their history of subjugation, a condition that equal rights would soon cure. Affirmations of female intelligence and ability, coupled with denials of female incapacity, featured prominently in feminist rhetoric. A common rhetorical maneuver based on this notion was to charge that women were unjustly categorized with others who were legitimately excluded from the democracy. British and American suffrage posters depicted women, perhaps wearing the gown of a college graduate or otherwise looking intelligent and capable, in the company of slope-browed, wild-eyed men of degenerate appearance, identified implicitly or explicitly as "idiots" and "lunatics." One such poster was captioned, "Women and her Political Peers." Another: "It's time I got out of this place. Where shall I find the key?" At a women's rights convention in 1867, the suffrage supporter George William Curtis asked why women should be classed with "idiots, lunatics, persons under guardianship and felons." Elizabeth Cady Stanton protested in 1869 that women were "thrust outside the pale of political consideration with minors, paupers, lunatics, traitors, [and] idiots."[17]

When historians have investigated the use of disability to deny women's rights, their objective has been to examine gender rather than disability. Lois Magner, for example, examined how women were said to bear the "onerous functions of the female," which incapacitated her for "active life" and produced a "mental disability that rendered women unfit" for political engagement. Nancy Woloch noted that a "major antisuffragist point was that women were physically, mentally, and emotionally incapable of duties associated with the vote. Lacking rationality and sound judgment, they suffered from 'logical infirmity of mind.' . . . Unable to withstand the pressure of political life, they would be prone to paroxysms of hysteria." Aileen Kraditor discussed how antisuffragists "described woman's physical constitution as too delicate to withstand the turbulence of political life. Her alleged weakness, nervousness, and proneness to fainting would certainly be out of place in polling booths and party conventions." On the one hand, this was of course an unfounded stereotype deserving of ridicule, as Kraditor's ironic tone suggests. On the other, just as it was left unchallenged at the time, historians have left unchallenged the notion that weakness, nervousness, or proneness to fainting might constitute a legitimate disqualification for suffrage.[18]

Decisions by immigration officials reflected the understanding that women had disabilities that necessitated their political, economic, and legal dependence upon men. An unmarried pregnant woman was doubly disabled, and when Maria Radke confronted immigration officials, her dependency was beyond all doubt. Radke was twenty-four years old and five months pregnant when she arrived at Ellis Island in 1906. Her brother, who lived in the far western part of the state, had sent her money for passage to come live with him. An official at her hearing asked, "Why didn't you compel the author of your condition to give you money or marry you?" Ignoring the part of the question about marriage, she responded that she had asked for money but been refused. Given that she was in what an official termed "a delicate condition," that she had "no one in the U.S. legally bound for her support," and that her father (whom she had not told about her pregnancy) was in Prussia, the board quickly decided, without debate or even discussion, that Radke was likely to become a public charge and should "be returned" to her father.[19] The case was unusual only in that immigration officials did not often have occasion to reject unmarried pregnant women; ticket agents and ship officers typically did so in advance.[20] Pregnancy was not requisite to being considered dependent, however. The immigration historian Martha Gardner found that regardless of work skills or experience, "women arriving during the early twentieth century who were alone, pregnant or with children, or with a checkered moral past were routinely found to be LPC [likely to become a public charge]." As Gardner nicely put it, "Poverty, in essence, was a gendered disease."[21]

Children

Children often presented a problem. Like women and many disabled people, they were dependent and admissible only when accompanied by an adult responsible for their care and support. The foremost question about immigrant children was whether their dependence was likely to be, in the normal way, a temporary condition. In spite of the clear and untroubled desire of many in the public and Congress to exclude disabled children, they posed difficult choices for immigration officials, who often ended up reluctantly admitting them when accompanied by nondisabled parents. The Immigration Bureau made a genuine effort to keep families together whenever possible, especially parents with their minor children. By admitting disabled children, they hoped to avoid confronting parents with a terrible choice between, on the one hand, returning as an intact family to the sometimes intolerable situation they had fled or, on the other, dividing the family by having one parent return with an inadmissible child while the rest remained. Still, officials permitted entry to children judged to be defective only if three conditions were met: that the parents were otherwise "desirable immigrants," that they were able and willing to post a bond guaranteeing their support for the child, and that a decision to exclude would divide a currently intact family. In addition, such "bond cases" had to be approved by the secretary in Washington.[22]

Immigration officials, however, often expressed frustration when they felt compelled to admit disabled children. For example, in 1913, when Keila Flacksman arrived in Boston with her deaf eight-year-old daughter, Feige, to join her husband and four other children already in the country, the commissioner at Boston worried that Feige was "likely to be a dependent for life and if placed in any school for her kind the family will probably endeavor to avoid paying her expenses." However, he reluctantly conceded, "I suppose under the circumstances the family should be landed if a bond is given." The commissioner general echoed his reluctance, but concluded that "under the circumstances there seems no alternative but to admit these aliens."[23]

When it was possible to reject disabled children without violating the policy on keeping families intact, officials usually did so. Sophie Fuko of Hungary arrived at Ellis Island in early December of 1912, with her six-year-old son, Kalman. Fuko's husband had died four years earlier. At the age of forty-six, with no remaining relatives in her native land, she had decided to join her two adult sons, Laszlo and Bela, in the United States. Upon arrival, however, the medical inspectors certified Sophie Fuko as "practically blind in [her] right eye" and her son as "afflicted with deaf mutism," and therefore both of them as "likely to become public charges." At their hearing before the Board

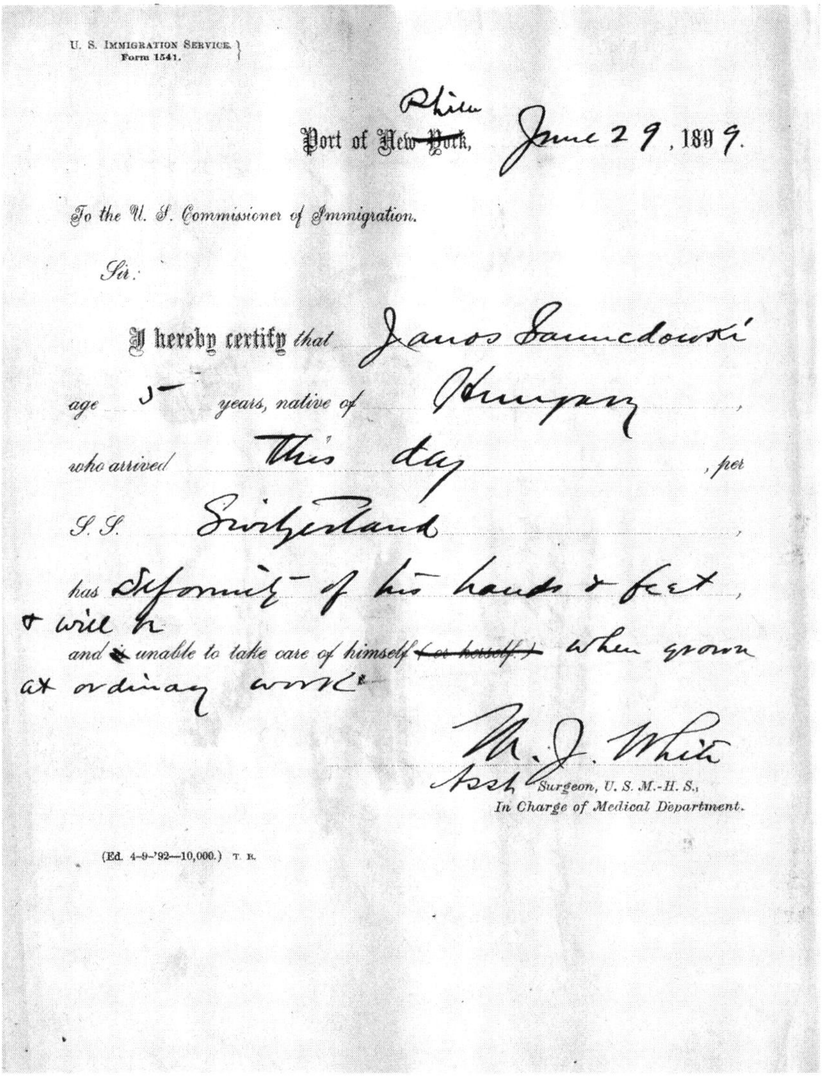

U. S. Immigration Service.
Form 1541.

Port of ~~New York~~ Phila, June 29, 1899.

To the U. S. Commissioner of Immigration.

Sir:

I hereby certify that Janos Samedowski

age 5 years, native of Hungary,

who arrived this day, per

S. S. Switzerland,

has deformity of his hands & feet,

& will be

and ~~is~~ unable to take care of himself ~~(or herself)~~ when grown

at ordinary work.

M. J. White

Asst. Surgeon, U. S. M.-H. S.,
In Charge of Medical Department.

(Ed. 4-9-'92—10,000.) T. R.

FIGURE 6. "I hereby certify that Janos Samedowski [?], age 5 years, native of Hungary, who arrived this day per S.S. Switzerland, has deformity of his hands and feet, and ~~is~~ will be unable to take care of himself (~~or herself~~) when grown at ordinary work." In the case of children, physicians were asked to make predictions of employment prospects years in the future. (National Archives, Records of the Immigration and Naturalization Service, Reports of Medical Inspectors in Philadelphia and New York, 1896–1903, RG 85, entry 2, vol. 1896–1900.)

of Special Inquiry, Fuko testified that she had always been self-supporting in Hungary as a housekeeper, while Laszlo and Bela gave evidence that they were both reliably employed and earning good wages. The family's finances and the presence of independent adult sons might have been sufficient despite Fuko's lack of a husband had it not been for the certificates. The board ruled that Fuko and her son were "suffering from physical defects, the nature of which will affect their ability to earn a living," and ordered them returned. Since impaired vision in one eye was not generally considered sufficient reason to exclude, Kalman's deafness was probably the decisive factor.

In her letter of appeal, Fuko argued that her sons were prepared to furnish bonds guaranteeing their support if needed and pleaded that the only family left to her now lived in the United States. She tried to minimize her son's deafness by claiming that her son could hear when spoken to loudly, that "of late he begain [*sic*] to talk very nicely," and that he also could read and write. Her instincts in this were correct; in other cases, the ability to speak, read, and write weighed in the favor of deaf immigrants. However, the commissioner at Ellis Island at the time, William Williams, had built a reputation as a strict enforcer of the immigration laws, especially concerning defects. He argued that "her child will always be physically defective, and it would be improper to admit merely because of the relatives here." The commissioner general agreed that "the Bureau does not think the mere presence here of two sons affords any good ground for admitting these physically defective aliens." Since the policy on keeping families intact usually applied only to minor children and their parents, not to grown children, the secretary dismissed the appeal and ordered Fuko and her son returned to an unknown fate in Hungary.[24]

The tendency to see education for deaf children as an act of charitable benevolence distinct from ordinary public education often created difficulties for parents like Fuko. Local public schools rarely offered accommodations for any kind of disability, and because in most areas there were insufficient numbers of deaf students to maintain separate schools locally, those who attended school at all usually went to centralized residential schools maintained by the states. Residents at these schools, because they lived away from home, were defined as public charges. Immigrant parents of deaf children were often pressed by officials to explain how they planned to pay for their children's education. In 1912 the father of nine-year-old Schie Budwicky, for example, was asked what he would do with his deaf son. "I will take care of him, send him to school and do everything for him possible," the father replied. "What kind of a school do you expect to send him to?" the official asked. "There is a school for deaf mutes in Trenton [the New Jersey State

Institution for the Deaf and Dumb] and I will send him to that school and I can well pay for him." "Is that school a private one or is it supported by public funds?" the official pressed him, and the father had to admit that he did not know. He was then asked if he was capable of posting a bond to guarantee that his son "will not be placed in an institution of any kind, nor that he will become a public charge?" Only because the father had the means to do this was Schie admitted.[25]

When deaf children were admitted into the country and subsequently became public charges in this manner, however, immigration officials found it difficult to simply deport them from communities where they had already established ties and acquired defenders. Sylvia Maginsky was six years old when she arrived with her mother in Seattle in January 1918. Because they were on their way to join her father, who was living in Brooklyn, New York, she was admitted. The next fall, however, the Department of Public Charities reported to the commissioner at Ellis Island that Maginsky was "now a public charge at the Institute for the Improved Instruction of Deaf Mutes in New York City." Under the law, any immigrant who became a public charge within five years of entry could be deported if it was "from causes not affirmatively shown to have arisen subsequent thereto." Since, according to the subsequent report filed by the Immigration Service, "this child is deaf and dumb and was a person likely to become a public charge at time of entry," a warrant was issued for her arrest.

The school was ordered to deliver Maginsky to Ellis Island for a deportation hearing. The school president, Felix H. Levy, however, protested directly to the Department of Labor that "the status of the children in our Institution is primarily educational and not charitable. We base this contention broadly upon the proposition that a deaf child is just as much entitled to an education at public expense as a normal child." A flurry of memos and letters ensued between the Bureau of Immigration, the Department of Labor, the State Board of Charities, and President Levy. In the end, the bureau decided to drop the matter quietly if the state board would agree to go along. The board's superintendent, Robert W. Hill, conceded that the "State Board of Charities does not desire to separate this child from her father and mother," but he wanted it made perfectly clear that her admission was an exception to policy and would "not be permitted to become a precedent for future admission of deaf mute persons who might arrive at our ports." The principle of free public education, he insisted, applied "only to the public or common schools of the state," and children "committed" to schools for the deaf were by definition public charges. Immigration officials must be more rigorous in denying entry to deaf children, he argued, for in most cases "the certainty that they will become public charges immediately leaves little room for question."[26]

Immigration inspectors generally shared this conviction. One expressed it explicitly in a moment of candor when nine-year-old Moische Kremen came before the board in August 1905, declaring that "the boy, Moische, who is deaf and dumb, is a public charge as he stands before this Board." The reason for this extraordinary statement, he explained, was that deaf children were very likely to be "placed in some charitable institution" (that is, sent to school). Moische was one of three children who had come from Russia in the care of their aunt, Rasche. Their father had come to the United States five years earlier with his three eldest children. Their mother remained in Russia with the two youngest, planning to follow later. At the hearing, the father testified that he was a tea salesman earning a respectable salary and had gone to great expense preparing three rooms for his children. Asked why his wife had not come, he described an oft-used strategy of working-class immigrants: "I thought to bring the older children first and they could take care of themselves and would earn something and then I would send for my wife and smaller children." What worked for other immigrants carried risk for parents of disabled children, however. As the family was already split, and would remain so whether they were admitted or excluded, they lost the protection of the policy against dividing families. Where the Kremens made their mistake was in sending Moische too soon. Had he waited and come last with his mother, the policy would have worked in his favor.

Moische's siblings were admitted, indicating there was no concern about the father's ability to support his children, but the board voted to exclude Moische as likely to become a public charge and ordered Rasche to accompany him back to Russia. In their appeal, the family's attorney contested the notion that Moische would become a public charge. The father and eldest son were earning good wages, he argued, the eldest daughter was ready to begin work, and the son Mordache was nearly ready to do so. As for the suggestion that the father would place Moische in an institution, given that the family was far from destitute, "we are not aware of any institution that will accept a child under these circumstances without pay, just to save the parents the expense of maintaining it." Even state-supported residential schools asked parents to contribute to their child's upkeep when they could afford it. Nevertheless, the appeal was unsuccessful. Too late, the mother's brother, a well-to-do restaurant owner in New York, offered to take responsibility for the boy until the mother arrived. He was tersely informed that the case had been decided several days previously. Moische had already been sent back to Russia along with his aunt.[27]

Disabled Adults

The presumption that deaf people were public charges per se was not limited to children. Although other immigrants over the age of twenty-one were generally treated as adults, the usually clear distinction in immigration policy between childhood and adulthood became blurred for deaf immigrants. In order to be admitted, a hearing relative or friend usually had to vouch for them and post a bond that insured against their becoming public charges. Although the skills, work history, or employability of a deaf immigrant might be considered mitigating factors, they were secondary considerations. Deafness was the master status, and it implied perpetual dependence.

A dispute within the Immigration Bureau over the case of Yankel Falikman, a twenty-two-year-old carpenter from Russia, highlights this ambiguity. Falikman arrived in Philadelphia in the company of his parents and three siblings in September 1913 in order to join a brother and two sisters already living there. The local board and commissioner decided that "the alien's case should be considered separately from that of the other members of the family, in view of the fact that he has reached his majority and has been self-supporting." His family was consequently admitted while he was held for deportation. The commissioner general, however, ruled that the board had acted wrongly and ought to have held the entire family together for deportation, reasoning that because of his "defects he is peculiarly dependent upon his parents and the other members of his family for care and protection, even though he may be able to earn his living as a carpenter." With Falikman's family already officially admitted, he saw no alternative to admitting him as well: "as they are now here the case seems to be one in which resort should be had to the bonding provision of the law." In this case, being seen as dependent in the end worked in Falikman's favor.[28]

Gedalie Rothstein, a twenty-eight-year-old tailor, had no such luck. Rothstein came to the United States in 1913 with his elder brother, Abram, to join his mother and three brothers who had been living and working in Philadelphia for more than ten years. His brothers were also tailors, one of whose employers supplied immigration officials with a written job offer for Rothstein. In spite of this, the board made their usual declaration that "he would find it extremely hard if not at all possible to support himself." His family's congressman wrote to assure the bureau that "there is no danger of [either brother] becoming a public charge as they will be given employment as soon as they land and their brother, Isadore Rothstein . . . is a property owner and earning good wages at his business as a tailor." The family's attorney reiterated that as Rothstein had a job waiting for him, his deafness "will not interfere in any manner with his earning a living," and that his family was willing and able to submit a bond for him. The commissioner general nevertheless declared that "there is little question as to the correctness of the excluding decision." Rothstein was excluded, and his brother likewise, because Rothstein, as a deaf man, was "in need of a guardian."[29]

The uncertain status of deaf people as competent adults was made manifest when boards asked their hearing family members, in some cases the minor children of deaf parents, to speak for them. Occasionally a deaf immigrant was both traveling alone and literate in English, in which case officials conducted the interview in writing. Most deaf immigrants, however, had no opportunity to testify on their own behalf. This surely put them at a disadvantage from the start by creating an immediate impression of incompetence and dependence. In 1904 eleven-year-old Mascheim Barash was questioned in lieu of his father, a thirty-six-year-old tailor. They were at first excluded by the board but later admitted when an elder son, who had been living in the United States for several years, came to the rescue and posted a bond.[30] In 1912 Perl Buchbinder's eighteen-year-old daughter had to testify in her stead. They were ordered excluded but were admitted on appeal when the father and husband, a mason who had been supporting two sons in Boston for four years, arrived to offer a bond.[31] In such cases as these, it may be that knowledge of the policy against dividing families led them to send part of the family to get established, the rest following later with the deaf member of the family.

Adrianus and Helena Boer were less fortunate. When they arrived at Ellis Island in 1905 with their sixteen-year-old daughter, Helene, both parents were certified for deafness. At their hearing, Helene was asked to speak for her parents—not as their interpreter but as their representative. Adrianus was forty-three years old and his wife, Helena, forty-seven. The board directed no questions to Helene's parents, nor did they ask her to consult with them. Instead, she was expected to think for them and to answer for them, in spite of the fact that both could read and write Dutch. This was doubly anomalous, since parents typically spoke for their children and men spoke for women; but disability trumped both gender and age in determining authority.

Helene may well have had prior experience acting as the voice for her parents in public, but she was unlikely ever to have experienced anything like what she faced now at Ellis Island. In contrast to the noisy tumult of their landing and initial processing, pressed on all sides by crowds of tired and bewildered immigrants, the hearing the following day would be focused and intense, with three officials sitting behind a long table asking questions whose significance she could only guess at. Helene must have known that she carried the future of her family in her hands, that what she did or did not say would determine whether they would continue on to a new life in America or be forced to return to the one they had left, minus the money expended on their journey and the possessions they had shed before leaving. In addition to her anxious uncertainty about the meaning of the questions and the effects of her answers, the questions came to her secondhand, translated from English to Dutch, and her answers likewise passed out of her control to the incomprehensible words of the interpreter, whom she had never met and did not know if she could trust.

Helene explained that her father was a skilled craftsman, a saddler and leather worker; that all members of the family were literate; and that they were going to join a family friend in Kalamazoo, Michigan. Her father, she said, had "always worked and supported the family without any outside help at all." Under immigration law, however, not one of them was an independent adult: two because they were defectives (the mother doubly dependent because both deaf and female), and one in part because of her age (although sixteen-year-old men on their own were sometimes admitted) but mainly because of her sex.

The board voted to exclude all three as "likely to become public charges." The Boers appealed the decision and might have succeeded, for their family friend in Kalamazoo had written to the Immigration Bureau explaining that he was an American citizen and willing to post bonds for all three, had Helene not then made a mistake. As part of the appeal process, the Ellis Island commissioner interviewed her personally. Perhaps because she felt safer and more trusting in his private office, perhaps because she thought it might help their case in some way, she confided that "my intended husband, Leopold De Boss, is expecting to come by the [S.S.] Rotterdam . . . I have made myself liable to be pregnant by him, but am not positive whether I am or not, as the time is too short to tell."

Here the issues of defect and gender became intertwined in complex ways. Mental defect was considered by experts at the time to be the root cause of much immoral behavior, while deafness as a condition was often thought to straddle the line between mental and physical defect, in many cases a result of the same degenerative conditions that lay behind mental and moral defects. As a child of deaf parents, Helene was already eugenically suspect. Now she had confessed to a significant moral lapse. Furthermore, pregnancy in itself was considered to be a disability that made women likely to be public charges. In some cases pregnant women were allowed to wed their intended husbands in ceremonies at Ellis Island and then to enter the country. This option was ruled out when Helene further confided that her intended husband was himself deaf. For the daughter of deaf parents to conceive a child with a deaf man would have been seen as an act of reckless irresponsibility.

All of these interconnected considerations were now brought into play: the subtle gradations of defectiveness that permitted some disabled people to enter while others were debarred; equally subtle gendered considerations that allowed certain women entry and denied it to others; and finally, complex and gendered moral distinctions. The appeal, with the damaging new information, was sent to Washington. Adrianus Boer wrote a short letter explaining, "I am a very skilled workman. . . . I am convinced that I will find work in Kalamazoo in a couple of days," and closing with a plea: "I pray you help us!" He did not address the matter of his daughter; perhaps he did not know. The Ellis Island commissioner wrote to Washington emphasizing the eugenic aspects of Helene's pregnancy, that she had become pregnant while still unmarried, and that she was a dependent of parents who were themselves dependents. The Boers' appeal was rejected and the three of them were returned to Europe.[32]

Occasionally a case arose where the assumption of dependency was so obviously misplaced that immigration officials had difficulty sustaining it. Charles McHardy, a deaf twenty-five-year-old Scottish stonecutter, arrived on Ellis Island on May 19, 1912. Because McHardy was both traveling alone and literate in English, he had the rare privilege of representing himself. As the commissioner wrote in his report, "The examination in this case was carried on by writing, the alien appearing very intelligent along this line." The transcript of the hearing nevertheless suggests a constant tension between McHardy's perception of himself as an independent adult and craftsman and the officials' difficulty in seeing him as anything but a dependent. When the board discovered that he had two brothers in Toronto, McHardy was asked twice whether he could not go to live with them. Thinking of himself first as a skilled craftsman, he replied that "the granite trade is not so good in Canada as it is in the United States." When he insisted that he did not "want to be dependent on anybody, as I can look after myself," an official responded by inquiring where his parents lived. In Scotland, was the puzzled answer. After he mentioned that one of the reasons he wanted to be in the United States was that he had many friends who now lived here—"nearly all are my old work fellows"—he was asked if any of them would be willing to post a bond for him. He protested that he "would rather not bother them as I know I will be capable in looking after myself." McHardy tried to focus attention on his skill and work history, pulling out his Masons' and Granite Cutters' Union card, but to no avail. Asked if he had any final statement as to why he should be admitted, his defiant reply was, "Because I am capable of work and I am a man." By this, of course, he meant that he was not a dependent—not disabled from work, not a woman, not a child. The board, seeing something else, declared him likely to become a public charge and excluded. Informed of his right to appeal and to apply for admission on a bond, he finally agreed to the role that had been scripted for him and telegraphed to a friend to act as his guarantor.

The Ellis Island commissioner's letter to Washington was ambivalent: "being deaf and dumb is probably not as serious a handicap in the trade of stone-cutter," he speculated, "as it would be in almost any other equally profitable trade." In fact, deaf people in America worked in many other equally profitable skilled trades, but the Immigration Bureau had never investigated the matter. Confronted with a skilled stonecutter for whom deafness had not been a handicap, the commissioner adjusted his stereotype just enough to accommodate the evidence in front of him without having to go any further. However, he quickly added, "of course, deafness increases the likelihood of being injured by accident whether at work or elsewhere." That is, as soon as one untested assumption was disrupted, he turned to another on which discrimination might be justified if required. He ended, "It seems to me that this is a very close case, but as the alien has no relatives here, I think deportation would not be much of a hardship."

The commissioner general was likewise ambivalent. Uncertain how to proceed, he wrote to the Granite Cutters International Association of America to ask whether "the fact that he is deaf and dumb will interfere seriously with his ability to obtain employment." The secretary of the association responded that, far from being a disadvantage in getting work, "whether it be from sympathy or because they will not waste much time in idle talk, employers seem to take kindly to them." They were superior workers, in fact, "their other short comings [*sic*] seemingly giving them additional powers in other directions." This endorsement of deaf workers' ability not only to do the work but to obtain employment was important. In other cases involving disability, officials made it clear that they took into account not simply their own assessment of an immigrant's ability to work, but also the possibility that they would face discrimination by employers.[33] With this encouragement, the commissioner general decided to recommend admission—but only with the bond that a friend and fellow granite cutter had offered to provide. The secretary approved, and McHardy was permitted to enter the United States.[34]

As the commissioner stated, it had been "a very close case." A successful, skilled worker in the prime of his life with no dependents had come to the United States to pursue a trade much in demand. The only way to exclude him was to pronounce him "likely to become a public charge," yet he was so manifestly *unlikely* to become one that there was no reasonable justification for exclusion whatsoever. Yet for every official involved it had been a close case. It could be so only because the presuppositions they had about the group to which McHardy belonged were so powerful as to nearly obliterate the individual they saw before them.

Although officials clearly saw the social class of immigrants as a legitimate factor to take into account when gauging immigrants' ability to support themselves, middle-class professionals who were disabled were nevertheless presumed dependent until proven otherwise. The burden fell on Albert and Evelyn Elliott, for example, traveling in the company of Frederick Baglow, to persuade officials that they would not become public charges. The three arrived from England at Ellis Island on February 5, 1906. Albert, at thirty-four, was a painter and a bookbinder. He carried over $100 in cash, deeds for property in England worth $1,250, and prepaid train tickets to Los Angeles, where they had friends. No occupation was given for his wife Evelyn. Their companion, Baglow, was a "sculptor and literary man" traveling as an "authorized Representative and Correspondent for 'The British Deaf Times,'" according to a letter he carried from the managing editor. He had a paid train ticket for New Orleans, where he was to stop over on his way to Los Angeles, and a check from his newspaper for $100.

The board looked at the evidence and concluded that "afflicted as they are . . . they would experience difficulty in becoming self supporting." Therefore, "having no one in the United States legally bound for their support or maintenance," all three were "likely to become public charges." Again the routine definition of the category to which they belonged overpowered the facts of the individual case and was sufficient for the board to refuse entry. The Elliotts and Baglow appealed the decision, noting that they were professionals and "have always supported ourselves." In this case, the tide was turned by the evaluation of Robert Watchorn, the Ellis Island commissioner, who was often criticized by restrictionists for insufficient zeal in excluding defectives during his time in office.[35] He wrote in his letter to Washington that "this case is of more than usual interest by reason of the fact that one of [the] appellants, Fredrick Baglow, although totally deaf is able to talk fairly well and understands questions asked of him by watching the lips of the speaker." He went on to note, "I was impressed with the excellent English which he used as well as by his good penmanship." Evelyn Elliott, he noted, "although certified as a deaf mute, has been taught to say a few words and can also understand what is said to her although in a less degree than Baglow." Such were the factors on which these immigrants' lives turned. He concluded that "the three appellants are so intelligent, well-dressed, ablebodied and anxious to work that I cannot believe there is any great likelihood of their becoming public charges." The commissioner general, after reading Watchorn's description, was convinced that these were "evidently people of education and refinement," and the secretary approved their appeal.[36]

Most immigrants were not, however, refined, well dressed, middle-class, or possessed of good penmanship. For both deaf and blind immigrants, stock phrases denoting dependency were the norm. Moische Kremen, the deaf child discussed earlier, was "helpless from the condition stated."[37] A twenty-eight-year-old dressmaker, Chave Unger, was described as "absolutely incapable of self-support, care or maintenance," although one of the board members offered in her defense that her uncle "offers to take care of her for the balance of her life."[38] Chajke Feinberg, twenty-three years of age and deaf, was "helpless from the cause certified."[39] Leopold Perelman, "being a dead [*sic*] mute, is a dependent" and "could not possible [*sic*] compete with an ordinary workman in this country."[40] Jane Potts, age thirty-four, was blind, "which, of course, makes her a dependent."[41] A seventeen-year-old blind immigrant, Esteban Martinez, was "helpless on account of his affliction."[42] In 1910 Karoline Jacobsen of Norway, thirty-nine years old and blind for the past twenty-five years, was asked, "Upon whom have you been depending for support?" She responded, "Depending upon myself for a living. I have been a knitter and besides I had a little money of my own." They then asked her brother, who came with her, "Have you contributed to her support in the past?" He responded, "No, she did not need any assistance." The board concluded that "she is helpless from Blindness" and that "we do not believe she will be capable of self-support, or of any material contribution towards same."[43]

Even travelers on temporary visits, if they had disabilities, had to produce proof that someone would take responsibility for them. In May 1905 two British missionaries, George and Anna Murray, who had been working in Palestine for the past fourteen years in the employ of the Christian Missionary Alliance of New York, were detained at Ellis Island. They had come at the invitation of the Alliance for a visit of five months to speak publicly about their experiences in the field. Although they remained on salary with the Alliance and were independent adults, and one board member voted to admit them, the other two voted to exclude. George was certified as having "deformed feet" and Anna as blind. No reason for exclusion was given outside of their respective disabilities. The letters included in their appeal all emphasized the wherewithal and responsibility not of the couple, but of their host. A letter from the missionary organization that had, in effect, become their guardian promised that "they are under the care of the Society while here" and that it would be "responsible for their departure." George Murray's letter reiterated that he and his wife would be "under their care and jurisdiction while in the United States." The Ellis Island commissioner wrote that he "was in grave doubt whether, in view of said defects they should be permitted to enter on their own representations"; however, an officer of the Alliance "so earnestly and fully assured me that his Society will consider itself morally and legally bound to care for them during their sojourn in the United States, and to provide for their departure . . . that I feel constrained to recommend that the appeal be sustained." In the end, the Murrays were allowed into the country—as dependents.[44]

On May 13, 1916, Robert Middlemiss, a sergeant major in the British army, was detained with his wife, Beatrice, at Ellis Island. Almost exactly a year earlier, Middlemiss had been blinded by a grenade on the second day of the doomed British campaign to capture Gallipoli. He had been invited to go on a speaking tour by George Kessler, a New York businessman who, while clinging to an oar for seven hours after going down with the *Lusitania* (coincidentally also almost exactly one year earlier), had determined to devote his considerable wealth, society connections, and business acumen to helping victims of war. Six months after his rescue at sea, he established the Permanent Blind Relief War Fund in New York City. One of the Fund's first public acts was to bring Middlemiss to the United States for a speaking tour. His topic was the ability of blinded veterans to work and live independently.

In March, two months before his arrival, a representative of the Fund telephoned the chief clerk at Ellis Island to inquire whether a blinded soldier would encounter any difficulties entering the United States. He then followed up with a letter to the commissioner to confirm the assurances of the clerk that "we would be safe in going ahead."[45] He wrote as well to the clerk to inform him that, "acting on your personal advise [*sic*], we will . . . complete arrangements," and to request that he personally follow up with the commissioner and then telephone the Fund for the purpose of "confirming the understanding." On May 10, the Fund representative wrote to Commissioner Frederic Howe, to inform him that Middlemiss would soon arrive. Aware of the kinds of objections that might be raised to admitting a blind man into the country, he took two approaches. One was to emphasize Middlemiss's competence as an independent adult, that "he is well-educated, has been trained in trades not requiring sight, is wholly capable of supporting himself." The other was to accept the definition of a blind man as a dependent, to reiterate that Middlemiss would be accompanied by his sighted wife, and to pledge that his "organization is ready to give all formal necessary official assurance." He ended by reminding Howe that "in causing Sergeant Middlemiss to come to this country under our auspices to lecture, we have acted on the informal and unofficial personal assurance from Mr. Sherman, of your office, that unless for some unforeseen condition, the man would be allowed to land."[46]

Middlemiss arrived two days later, was duly certified by Surgeon Turnipseed as "BLIND IN BOTH EYES which affects ability to earn a living," and sent before Inspectors Toner, Harper, and Bruno of the Board of Special Inquiry. Middlemiss emphasized his own competence, stating that he had trained as a masseur at a London hospital after his injury, had in his possession $175, and had a pension from the British Army. He added that their three-year-old daughter was being looked after by Beatrice's mother while he and his wife were away. None of this suggested that he might try to remain in the country permanently, or that he would become a public charge if he did. Osmond Fraenkel, an executive committee member for the Permanent Blind Relief War Fund, testified next and attempted to head off any question of Middlemiss's becoming a public charge. He stressed that "despite the alien's blindness he is able to earn his living should such a contingency arise," and at the same time promised that Middlemiss "will be taken care of by the persons who invited him to come," and "will in no way become a burden here."

The object of Middlemiss's visit, Fraenkel continued, was to speak on how blinded soldiers can become "able to make a living and become self-supporting and hopeful citizens." This was not a message the board was prepared to consider, however, and it concluded that Middlemiss "in his present condition would be incapable of self-support, and his wife is dependent upon him for support." The assurances of employability were beside the point, for he was blind. Beatrice was a trained nurse, but under immigration law she was, primarily, a wife. The testimony of their bodies was that both were dependents. The board followed law, regulation, and common understanding, pronouncing both "persons likely to become a public charge."

That same day George Kessler wrote personally to the Secretary of Labor requesting (or rather politely demanding) that the board's decision be overruled immediately. Near the start of the letter he pointedly noted that the Fund was "directed by a number of American gentlemen of wealth, standing and good repute," and in his penultimate paragraph that the "organization desires to call your attention to the names of the persons of standing and substance who are officials of it, and whose names are on the accompanying letterhead." Kessler reminded the secretary of the informal assurances given by an Ellis Island official and assured him that the Fund or he personally would be willing to finance a bond ensuring that "neither Middlemiss nor his wife will become public charges." In closing, he was "trusting that you will grant our appeal at the earliest possible moment and will admit Mr. Middlemiss and his wife by instructions to the Ellis Island authorities."[47] That letter was dated May 12, the same day of Middlemiss's arrival and hearing. The very next day, before the commissioner could mail his own recommendation on the case to Washington, "telephonic instructions" arrived from Washington ordering the immediate admission of the Middlemisses to the United States.[48] Middlemiss went on his speaking tour, at least in part because a wealthy and influential sponsor was there to act in loco parentis for the thirty-five-year-old sergeant major.

Three years later Erik Harildstad arrived from Norway at Ellis Island on his way to study pedagogy at the Perkins Institution for the Blind in Boston. He had a nine-month fellowship for this purpose from the American Scandinavian Foundation of New York City, which had arranged lodgings for him. He carried with him his teacher's certificate and a letter of introduction from the Royal Norwegian Department of Public Worship and Education. He was "found to be afflicted with loss of sight of both eyes" and sent before the Board of Special Inquiry.[49] A foundation representative testified and produced letters from the foundation following the same two-track defense that had been used for Middlemiss. Harildstad had graduated with honors from the school of education at Elverum, Norway, was employed as a teacher at the Educational Institute for Blind in Christiania and was the president of the Norwegian Association for the Blind and the editor of its newspaper. Furthermore, he had translated several books from English and published several articles on education for the blind. At the same time, the representative furnished the board with a document that committed the American Scandinavian Foundation to full responsibility for his care and support while in America. Following this evidence came the decision: "In view of the grave nature of his certified condition, the Board is of the opinion that he is likely to become a public charge. Excluded and ordered deported."[50]

Harildstad was an unusual case—a professional with a college degree, steady employment, and a guarantee of full support while temporarily residing in the United States. He was blind and therefore presumed dependent but, like Middlemiss, his sponsor was well connected. Although occasionally a member of the House of Representatives would intervene on behalf of an immigrant who had family in his district, senators did not often do so. In this case, however, Senator Knute Nelson of Minnesota, who had immigrated to the United States from Norway with his family as a seven-year-old, seventy years earlier, protested to the commissioner general the following day that "an injustice is being done" to a man who "could in no way be a public charge."[51] The Norwegian minister wrote to the Secretary of State asking for his assistance, giving assurances that both the consul general and the foundation had "promised to take care of Mr. Harildstad while he is in the United

States and provide for his return to Norway."[52] The American Scandinavian Foundation sent supporting materials to Washington, taking pains to point out that "the Plan of Studies inclosed [*sic*] was typewritten by Mr. Harildstad himself" (it seems unlikely that such a statement would ever be made for someone not a child, or not disabled).[53]

The Secretary of State then wrote to the Secretary of Labor to inform him that the Norwegian legation had agreed "to hold Mr. Harildstad ready for deportation at any time during his stay in this country should conditions arise making it necessary or advisable,"[54] and then wrote again two days later to assure him that he had a signed letter "from the Norwegian Minister, guaranteeing that Mr. Harildstad will be cared for during his stay . . . and that he will depart again after the termination of his studies."[55] The commissioner at Ellis Island supported the board's decision, writing that Harildstad "has no relatives in this country, and in view of his affliction, I do not feel that he should be admitted, even temporarily, unless an enforcible [*sic*] bond is filed in his behalf" with a firm requirement that he leave the country within one year. The commissioner general, however, noted that "alien is an accomplished and highly educated person. He is coming here to study pedagogy, psychology and the higher branches of the English language. He can read with his fingers and is able to type on the typewriter." (Again, this is likely the only memorandum from the Secretary describing an "accomplished and highly educated" immigrant where he saw fit to specifically point out an ability to read and to use a typewriter. But then disability is often translated into a generalized inability, and demonstrations of competence, even at relatively simple tasks, often elicit astonishment.) His recommendation was that Harildstad be "TEMPORARILY ADMITTED . . . upon the filing of a public charge bond," and the next day, the Secretary of Labor upheld the appeal.[56] It took ten days; letters from a prominent foundation, the Norwegian consul, the Norwegian minister, a senator, and the Secretary of State; and a bond that guaranteed he would not become a public charge to get a blind man into the country for a short visit. Harildstad departed, as promised, the following July.[57]

Less typical professionals required similar guarantees. Gundia Roe, also known as the "monkey-man," and Supermoney Munsammy arrived from India under contract with the Barnum and Bailey and Ringling Brothers circus. They were either ten and eleven years old, respectively, or fifteen and nineteen, depending on which document in their immigration file is correct. Roe was "certified as an idiot" and refused entry on that basis. On Munsammy's certificate the physician wrote, "Dwarf and curvature of the spinal column." To the board he identified himself as "a little midget-man—a showman" who had toured with the circus in the United States three times in the past. When asked what he had done for the circus during his previous visits, he responded, "I just sat there and the lecturer lectured about me." This was not the right answer, for although the law provided exemptions that allowed performing artists on tour to enter the country, it was silent about people who just sat to be looked at. The board decided that Munsammy "is not an artist; he is a freak of nature, coming here solely for exhibition purposes," and excluded him as likely to become a public charge. John Ringling himself wrote a letter of appeal. Both "can be considered as artists," he explained, "as they are able to and do perform dances characteristic of their native lands." Roe and Munsammy would be "under our direct control, with a separate attendant for each individual, night and day, during their entire stay in this country . . . at no time will they be permitted to mingle with the general public." Commissioner Williams found this a satisfactory arrangement for Munsammy, accepted a "broad meaning" for the word "artist," and recommended temporary admission on a $1,000 bond. But Roe was "certified to be an idiot," making it "impossible to allow him to land under any condition." Acting commissioner general Frank Larned agreed that Roe was "mandatorily excluded and has no appeal." For Secretary Nagel, however, few decisions were mandatory or impossible. Exercising his discretion as a political appointee and his responsibility to look after prominent constituents, he simply penned in at the bottom of Larned's memorandum, "I think they may both be admitted." Roe and Munsammy toured for eight months before they were returned to India, as guaranteed by Ringling.[58]

Antonio Maltese regrettably had no influential sponsors. He was blind in one eye and had vision of 10/100 in the other, according to the certificate made against him in April 1920. He had a brother-in-law in Detroit, which in Italy might be thought to entail obligations second only to parenthood, but in the view of the bureau Maltese had "no one in this country legally or morally bound to assist him." His brother-in-law forwarded a bond affidavit pledging to support Maltese if needed and described his own finances: eight years as a machinist at the Detroit Creamery Company at $45 a week, $600 saved, a house worth $10,000. Attached to the affidavit was $25 for travel money. Maltese was asked what he would do in Detroit: "I will making macaronee [*sic*]," as he had done in Italy, he told the board. When asked about his "eye trouble," he responded simply, "I was born that way." It was a straightforward matter for the board too, which quickly agreed that he should be excluded as a likely public charge and asked Maltese if he wished to make any further statement. "I am to work as a macaronee maker" was all he could think to say, and probably all that he thought necessary. The commissioner, describing

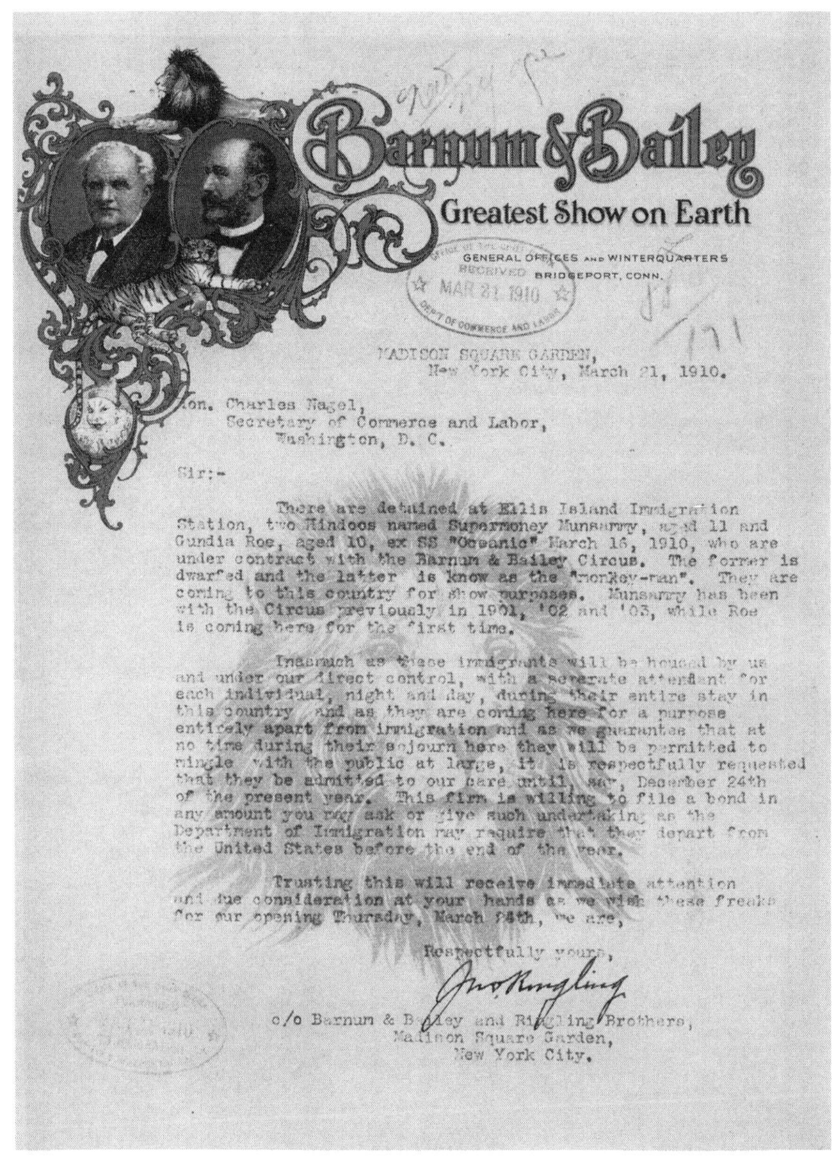

Barnum & Bailey
Greatest Show on Earth
GENERAL OFFICES AND WINTERQUARTERS
BRIDGEPORT, CONN.

MADISON SQUARE GARDEN,
New York City, March 21, 1910.

Hon. Charles Nagel,
Secretary of Commerce and Labor,
Washington, D. C.

Sir:-

There are detained at Ellis Island Immigration Station, two Hindoos named Supermoney Munsamy, aged 11 and Gundia Roe, aged 10, ex SS "Oceanic" March 16, 1910, who are under contract with the Barnum & Bailey Circus. The former is dwarfed and the latter is know as the "monkey-man". They are coming to this country for show purposes. Munsamy has been with the Circus previously in 1901, '02 and '03, while Roe is coming here for the first time.

Inasmuch as these immigrants will be housed by us and under our direct control, with a separate attendant for each individual, night and day, during their entire stay in this country and as they are coming here for a purpose entirely apart from immigration and as we guarantee that at no time during their sojourn here they will be permitted to mingle with the public at large, it is respectfully requested that they be admitted to our care until, say, December 24th of the present year. This firm is willing to file a bond in any amount you may ask or give such undertaking as the Department of Immigration may require that they depart from the United States before the end of the year.

Trusting this will receive immediate attention and due consideration at your hands as we wish these freaks for our opening Thursday, March 24th, we are,

Respectfully yours,

Jno. Ringling

c/o Barnum & Bailey and Ringling Brothers,
Madison Square Garden,
New York City.

FIGURE 7. Letter from John Ringling of Barnum and Bailey and Ringling Brothers to Charles Nagel, Secretary of Commerce and Labor, March 21, 1910. (National Archives, Records of the Immigration and Naturalization Service, RG 85, Accession 60A600, file 52880/171.)

him as "this 34 year old Italian macaroni-maker," tersely concluded that "his certified physical condition and the other facts set forth in the record amply justify exclusion." The commissioner general's memorandum, headed "In re Antonio Matese, aged 34, native of Italy, Italian race," likewise recommended against admittance. The Secretary concurred without comment.[59]

When immigrants were rejected, they usually left the stage of history and disappeared from the historical record. Moische Fischmann, whose story began this chapter, did not. Unlike most rejected immigrants, several years later Fischmann had a second chance. When he was rejected as likely to become a public charge, rather than return to Russia he traveled to Antwerp, Belgium, where he found steady employment until the outbreak of World War I. During the siege of Antwerp by the German army, he was employed by the Belgian government to dig trenches. When Antwerp fell in October 1914, he fled together with other refugees, eventually ending up in London, where he went to work as a laborer in one of the refugee camps established by the British government. A local philanthropist active in refugee work took an interest in Fischmann and, after corresponding with his family in America and learning of their desire to have him with them, paid for his passage to New York. On November 11, 1915, Fischmann found himself again in the inspection line at Ellis Island.

Again he was certified by the inspectors as "afflicted with a physical defect" and sent before the Board. There a similar scene played itself out: Fischmann's brother testified—now a bit more prosperous, employed as an ironworker at $22 a week and having invested in four real-estate lots. His sister had since married, and in accordance with custom her husband testified on her behalf. He was a dress presser earning $25 a week and was willing to teach Fischmann his trade if, by some chance, no jobs for ironworkers were available. Two cousins promised whatever help was necessary. Again the board excluded him as "likely to become a public charge" entirely on the basis of his deafness.

Fischmann again appealed to Washington. Again, the Hebrew Sheltering and Immigrant Aid Society wrote in support, arguing that "we fail to find how that affliction in any way interferes with a person's ability to earn a living. There are thousands of deaf mutes in the United States and yet almost all of them are able to support and maintain themselves." As the Ellis Island commissioner pointed out, nothing material had changed in the case, except this—Europe was at war. Both the transatlantic journey and the destination were perilous. It was not feasible to return Fischmann. Few immigrants were coming to the United States during these years, and none of the few who did arrive were forced to return. In cases where exclusion was mandatory, as with mental disability, bonds were filed requiring that aliens present themselves for deportation once the war ended. In many other cases, such as this one, the Bureau simply let them in. The commissioner was chagrined that Fischmann's "admission will be forced by reason of the abnormal conditions prevailing abroad." The acting commissioner general was more philosophical about it, concluding that "it is very natural the relatives here were desirous of getting [the] alien out of London where his condition has doubtless been rather precarious. The Bureau believes it would be proper to admit this physically defective alien, whose nearest relatives are the brother and sister to whom he is destined, under bond, and it so recommends." Moische Fischmann made it.[60]

4

Ugly

When Harry Haiselden, an early twentieth-century physician and eugenicist, defended his euthanasia of a disabled infant, he explained that "it was terribly ugly." A "defective" such as this child "is not a beautiful thing," he wrote; "it is a monstrosity. It is not to be saved."[1] When the poet Amy Lowell wanted to criticize the writing of Randolph Bourne, she concluded her assessment with "deformed body, deformed mind."[2] When Robert Watchorn, the commissioner at Ellis Island, wanted to make clear why an Armenian refugee from Turkey should not be admitted into the United States, he stressed that the man was "repulsive in appearance."[3]

Appearance can figure just as importantly as function in the creation of disability, in some instances even more so. Prejudice against persons with functional impairments may be magnified by their anomalous bodies or behaviors; in some cases, prejudice may spring entirely from a sense of repulsion, as mild as unease or as strong as disgust. It is not uncommon for people to treat wheelchair users as less competent, a psychological parallel to the physical experience of talking down to them.[4] Persons with speech impediments may be thought less intelligent because they are less intelligible or "normal" sounding. People who have birthmarks, scars, or other impairments of appearance may have no functional impairment whatsoever but still be considered disabled and face discrimination.[5] Rosemarie Garland Thomson explains that in any first encounter, "a tremendous amount of information must be organized and interpreted simultaneously," and each person "prepares a response that is guided by many cues, both subtle and obvious." When one of them has a visible disability, it tends to overpower other cues, making the interaction "strained because the nondisabled person may feel fear, pity, fascination, repulsion, or merely surprise, none of which is expressible according to social protocol." That discomfort can incline nondisabled people to avoid further such encounters. Moreover, the disability often "cancels out other qualities, reducing the complex person to a single attribute."[6] In sociological terms, disability can become a master status, swamping other indicators of identity and social status.[7]

Psychologists began writing in the 1960s about the tendency of the nondisabled to avoid close proximity with the visibly disabled.[8] The political scientist Harlan Hahn later described the source of this discomfort, when confronted with unusual and stigmatized physical characteristics, as "aesthetic anxiety."[9] More recently, social scientists have connected that aversion to the sense of disgust, which they suggest may be an evolved mechanism for disease avoidance. That mechanism is unfortunately not finely tuned enough to reliably distinguish between dangerous contagious diseases and harmless anomalous appearances. For example, "humans, and probably animals too, seem disposed to infer even innocent cues as possible signs of disease," a propensity often "seen in the stigmatization of people with facial abnormalities wrought by injury or non-infectious illnesses such as psoriasis, eczema or acne."[10] In one experiment, subjects were more willing to keep "a close personal distance with confederates who appeared 'normal' or who feigned a temporary condition (e.g. a broken arm) relative to those confederates feigning more permanent conditions such as an amputated leg or clubfoot." Another found that they "stood further away from people described as amputees or epileptics relative to people described as 'normal.'"[11] Further studies have found that people who generally experience pronounced anxiety about health and disease are more likely to avoid association with people with disabilities.[12]

A blunt disease-avoidance mechanism produced by natural selection may explain in part this tendency to avoid proximity to persons with disabilities, but a significant cultural element undoubtedly contributes as well. As Martha Nussbaum has observed, disgust felt toward primary objects such as feces, blood, and other bodily fluids, certain insects, or decaying meat is different from what she terms "projective disgust," a deliberate linkage of primary objects of disgust with stigmatized others. The latter is learned. Jews, African Americans, Roma, and "untouchable" castes in India, for example, have all been stigmatized in various ways by association with vermin, filth, decay, and disease.[13] The stigma of disability changes over time, waxes and wanes, and varies from place to place and culture to culture. It certainly exists today, yet it is far less intense than it was a hundred years ago. Aversion to disabled people is at least in part learned behavior, a function of familiarity and cultural conditioning.

Disgust can also result when the perceived dividing line between humans and other animals is blurred or threatened. Disease or disability may be experienced as disgusting because it conveys reminders of humanity's essentially animal nature. Nussbaum writes that "disgust has been used throughout history to exclude and marginalize groups or people who come to embody the dominant group's fear and loathing of its own animality and mortality."[14] The common aversion to thinking about disability, the philosopher Alasdair MacIntyre argues, arises in part from the "failure or refusal to acknowledge adequately the bodily dimensions of our existence" and the desire to "imagine ourselves as other than animal."[15] The late nineteenth century was a time of acute sensitivity about the question of humanity's animal nature. The most prominent and talked-about findings of science were those that demonstrated a close relation of humans to other animals. If disability was a reminder of the animal nature of humans, of their physical frailty and the inevitability of decay and death, it is not surprising that segregation, rejection, and fear of the disabled should have grown at the same time.

Another powerful source of feelings of disgust identified by Nussbaum is the fear that contaminants may cross the boundaries of the body. It is suggestive how often animus toward disabled immigrants was expressed in bodily metaphors. Public Health Service physicians described the immigration of defectives as a "crime against the body politic."[16] George Lydston, a professor of criminal anthropology, believed Asian immigrants to be "much less dangerous" than European defectives, who had become "veritable fungi on the American body social."[17] Charles Davenport called for an alliance of biologists and legislators to "purify our body politic of the feeble-minded."[18] James Davis, the Secretary of Labor, warned that "a nation's mind can be poisoned as soon as a nation's body is allowed to get soft," and that the "only biological way for a nation's body to become weak is for it to cease breeding from good stock."[19] In the *Saturday Evening Post*, an immigrant community was a "wriggling cross-section of life" that was becoming "imbedded in the fair physical corpus of New York."[20]

Ingestion was a popular metaphor for expressing disgust about the absorption of immigrants into the national body. In 1896 Francis Walker wrote that it was not possible "that we can take such a load upon the national stomach without a failure of assimilation, and without great danger to the health and life of the nation." The nation "should take a rest" to give the "system some chance to recuperate." Congressman Percy Quin of Mississippi declared in 1914 that the "United States has had so much of this scum that we have a bad case of colic" and will soon develop "acute indigestion." Making his point even more colorfully, he added that "the Government must belch, and where is it going to belch them? You can not get rid of them after they are once here."[21] A *Saturday Evening Post* writer fretted that "America is sick from trying to digest indigestible immigration, and she needs expert medical advice." Another feared what would come of the "national indigestion" caused by the "millions of undigested aliens" who were "hopelessly inferior in physique, manner of thought and ability."[22] French Strother, editor of *The World's Work*, thought that "the stomach of the body politic" was "filled to bursting with peoples swallowed whole whom our digestive juices do not digest."[23] A *North American Review* writer worried, "We have begun to gag a bit over the size and quality of the dose" of the foreign influx, which might "contain poisons against which we have no antidote."[24]

Defective immigrants were similarly often described as waste products and garbage, or, in Robert DeCourcy Ward's words, "'drainage' of the inland regions of Europe."[25] Lydston decried the "degenerate flotsam and jetsam" that had "entered our country in a continuous stream," treating it as "a dumping ground for the sweepings of Europe."[26] Irving Fisher maintained that the country had become "a dumping ground for relieving Europe of its burden of defectives, delinquents and dependents."[27] William Williams was exceptionally prolific on this topic, condemning the "invasion of physically defective foreigners" who "herd together in unsanitary surroundings," the "mere scum or refuse, persons whom no country could possibly want."[28] *Current Opinion* editorialized against the "stream of impurity," the "tide of pollution," the "turgid stream," and the "dumping of Europe's human refuse at our doors."[29] The journal *Medical Record* conjured up the oddest image of all in the editorial "The Invasion of the Unfit" when it described mentally defective immigrants as "lumps of poisonous leaven."[30]

Disgust existed not only in the background of immigration law, but on the face of it well: the so-called Class A categories, for which exclusion was mandatory, included not only "idiots, imbeciles, the feeble-minded, the epileptics, the insane," but also those with "loathsome or dangerous contagious diseases." The meaning of "dangerous" was reasonably straightforward, but "loathsome" diseases were ambiguously defined as "those whose presence excites abhorrence in others, and which are essentially chronic." Such diseases need not threaten death, serious harm, or contagion, but merely arouse disgust. Certain conditions were specifically included in that category, such as favid, ringworm, parasitic fungus diseases, leprosy, and venereal disease, but it was an elastic category.[31] The chief medical officer at Ellis Island in 1917 conceded that he knew of "no strictly medical classification or enumeration of loathsome diseases," and that whether or not "a disease is loathsome depends entirely upon the personal attitude of the observer, and that attitude may be largely influenced by his familiarity and association with the particular disease under observation."[32]

The anthropologist Mary Douglas once noted that the "more the social situation exerts pressure on persons involved in it, the more the social demand for conformity tends to be expressed by a demand for physical control." Douglas was referring mainly to the felt need to hide normal "organic functions" from public view, but it applies as well to the disruptive or disturbing bodily configurations, movements, or behaviors of the disabled.[33] The violation of such expectations as that people rise from their chairs at appropriate moments, confine the motions of their arms and legs to a limited range of acceptable possibilities, adjust their facial expressions to correspond to the situation, and manage their emotions to conform to a circumscribed palette of expression often triggers unease and avoidance. Limbs that seem out of control, faces that contort, and bodies that do not look or move as they are supposed to not only violate norms but also remind others of their own organic functions and frail physical nature, that their own bodies will sooner or later malfunction, decay, and die. The consoling image of an immaterial self within, apart and eternal, may wall off awareness of vulnerability and mortality, but the disabled body always threatens to bring the wall down.

The City

The turn of the twentieth century was a time of an exceptionally powerful "social demand for conformity." John Kasson attributes the increasing emphasis on conformity of appearance mainly to the influence of urbanization. In 1820 only about 7 percent of the population lived in urban areas, defined by the Census Bureau as communities of more than 2,500. The largest city in the country was New York, with all of 125,000 people. Philadelphia had 64,000, and Chicago barely existed. The 93 percent who lived in rural communities rarely encountered anyone whose occupation, family, history, virtues and vices were not well-known to them. By 1900, 40 percent of the population lived in urban areas, and by 1920 it was more than half as city populations swelled due to both immigration and internal migration from the countryside. From 1860 to 1910 the population of New York City quadrupled, and that of Chicago increased from a few hundred to over two million. It was much the same across the country, with dramatic growth occurring not just in the great cities but in small and mid-sized ones as well.

The inhabitants of rapidly expanding towns and cities found themselves making their daily rounds among strangers, on sidewalks, streetcars, and buses, in shops, elevators, and public lavatories. The growing anonymity and mobility of urbanizing America made personal appearance simultaneously more important and more subject to manipulation. Navigating the city demanded skill at interpreting clues in the appearance of others, in their bodies, faces, and clothing, to know whom to trust, whom to avoid, and who ranked higher or lower in relation to oneself. Visual inspection, interpretation, and evaluation of other human beings became a pervasive aspect of everyday life in the city, in the same way that reading weather was for farmers, or reading flora and fauna was for a foraging people. With everyone reading everyone else, conscious management of personal appearance became essential and habitual for anyone seeking to maintain or improve their status. Constantly on display, with no way of knowing when a casual encounter might have social consequences, the management of impressions became an invaluable skill. City dwellers needed to be prepared to prove themselves to strangers at any time and had less leeway for making varied impressions or letting down their guard. Most of those they encountered would have no history of interactions with them, or general knowledge of their character, to provide context and background. As opportunities for missteps and embarrassment in public multiplied, urbanites developed means to blend into their environment, akin to the camouflage of prey animals and plants—to reduce the chances of standing out, attracting attention, and risking embarrassment. They donned standardized attire for going out in public, avoided eye contact, matched their pace to the crowd, and if they stumbled, literally or figuratively, they recovered as unobtrusively as possible.[34]

All of this is second nature now to the majority of humanity, accustomed to living among strangers, but at the time many people found themselves at a loss in this rapidly changing social landscape—those newly risen into the middle class, newcomers to the city, the no-longer-young baffled by changing norms, and migrants from the countryside. As Kasson has shown, advice books proliferated to instruct anxious readers that, for example, wearing black was the safest and least eye-catching choice, or to counsel that in "large cities, men rarely walk in the street in their dress-suits without wearing a very thin overcoat, even in summer," and that readers had best conform if they wished to "avoid being conspicuous." As for women, "singularity is to be avoided, and she is best dressed whose costume presents an agreeable whole, without anything that can be remarked." There were warnings against scratching, ear boring, and fondling the mustache. "Don't beat a tattoo with your foot in company or anywhere, to the annoyance of others," one instructed, and "don't drum with your fingers on chair, table or window-pane. Don't hum a tune. The instinct for making noise is a survival of savagery."

Standardized speech should be cultivated, for a "strong local accent . . . marks you as underbred." [35] Kathy Peiss has similarly argued that a "fundamental change was taking place: the heightened importance of image making and performance in everyday life." She found that for urban women, the use of cosmetics to enhance beauty and disguise blemishes became an everyday necessity as they "promenaded the streets, looked in shop windows, socialized in public, and scrutinized one another."[36]

Standards of etiquette that demanded greater control over the movements of the face and body were expressed in advice to Italian immigrants to "try not to gesticulate, and do not get excited in your discussions" (the importance the authors attached to this advice might be suggested by what came just before it: "throw away all weapons you may have").[37] Americans in general were advised to "indulge in no facial contortions" and to "practice talking without moving the facial muscles but slightly," and to do this "before the mirror daily." They were encouraged to "practice keeping the arms pressed lightly against the sides in walking and sitting" and informed that "gesticulation in conversation is always in bad taste." [38]

What did this mean for disabled people? Individual experiences varied tremendously, of course, according to circumstances, but there are many indications of an increased stigmatization of disabled people based in part on a growing sensitivity and aversion to abnormal appearance. These demands for greater control over the face and body, for example, were expressed in a movement to forbid the use of sign language in schools for the deaf, lest the students look like apes, savages, or madmen. Critics condemned sign language as nothing more than "a set of monkey-like grimaces and antics"; teachers who used it in the classroom "grimace and gesticulate and jump." Students were told, "You look like monkeys when you make signs."[39] Even schools that continued to allow sign language complained of students' "indulging in the horrible grimaces some of them do."[40] A contributor to the magazine *Science*, describing a visit to a school for deaf students, was appalled by the "inmates making faces, throwing their hands and arms up and down," adding that the "effect is as if a sane man were suddenly put amidst a crowd of lunatics."[41]

According to Sander Gilman, exaggerated facial expressions were often described as a marker of atavistic insanity: "the absence of civilized standards of behavior" indicated "a return to earlier modes of uncontrolled expression."[42] Allan McLane Hamilton, "consulting physician to the insane asylums of New York City," in 1883 published an illustrated guide to the visible signs of insanity. Hamilton opened by noting that "when one walks through the wards of any asylum for the insane, he will be immediately impressed with the repulsiveness of the faces about him . . . plainness or downright homeliness is the rule among asylum patients, whether of high or low social station."[43] Philip Ferguson, in his history of intellectual disability, argued that a new "cultural aesthetic" and a "heightened attention to appearance" in the second half of the nineteenth century intensified the stigma associated with disability: "Ugliness became as important a judgment as beauty" but was not confined to the realm of aesthetics, for ugliness could also lead to "accusations of moral and mental failure."[44] Martin Pernick argues persuasively that aesthetics were at the heart of the rise of the eugenics movement of the early twentieth century: "Eugenics promised to make humanity not just strong and smart, but beautiful as well. Being hereditarily fit included being visually attractive. Ugliness was a hereditary disease." In the voluminous literature of eugenics, beauty was regularly equated with evolutionary fitness and ugliness with defect and degeneration.[45]

As the household became less important as a site of production over the course of the nineteenth century, the ideal middle-class home became a refuge from the competitive world of work and business, a "haven in a heartless world," in Christopher Lasch's words. It was a refuge not only from the world of work and business, however, but also from the eyes of strangers. As the pressures of living among strangers mounted in urbanizing settings, the need for a radical separation, a place to escape from the gaze of others, mounted as well. The home, for the middle class, became increasingly a private space where vigilance over the functions and malfunctions of the body might be relaxed. For middle-class family members with visible disabilities, the home also became a place of refuge from contemptuous or pitying eyes.[46]

For the poor, however, simply remaining at home was often not an option, and growing numbers of disabled beggars took to the city's streets. Susan Schweik has described how, in response, local governments enacted "unsightly beggar ordinances" to spare passers-by the sight of bodies that aroused disgust. San Francisco was an early pioneer in 1867 when it made it unlawful for "any person who is diseased, maimed, mutilated, or in any way deformed, so as to be an unsightly or disgusting object," to "expose himself or herself to public view." Portland, Oregon, added "crippled" to the list of unlawful beggars when it passed its version in 1881. Schweik discovered eight additional cities that followed with their own similarly worded laws: Chicago in 1881, Denver in 1886, Lincoln in 1889, Columbus in 1894, Omaha in 1890, and Reno in 1905. The Pennsylvania legislature passed a statewide law in 1891 prohibiting "the exhibition of physical and mental deformities . . . for hire or for the purpose of soliciting alms."[47] Rosemarie Garland Thomson described these laws as one expression of "a kind of bowdlerizing of the body" that took place in the late nineteenth and early twentieth centuries.[48]

For the working-class disabled, difficulty obtaining jobs might keep them at home, or it might push them onto the streets or into public institutions. The industrial-efficiency experts Frank and Lillian Gilbreth argued in 1920 that a majority of the public felt that having disabled people in the workplace was "repellent, that these people should be provided for by pensions in their homes."[49] Unfortunately, except for disabled war veterans, few people could obtain pensions. Disabling industrial accidents occurred with alarming frequency, but most injured workers had little recourse before states began to enact worker's compensation laws in the 1910s and 1920s, and these did nothing for those whose disabilities were unrelated to their employment. A steadily growing number of disabled people, for the most part those with psychiatric or intellectual disabilities, were institutionalized in the late nineteenth and early twentieth centuries. Families who could afford it were increasingly likely to place difficult or burdensome relatives in private institutions. The poor turned to the rapidly expanding public institutions, as did civil authorities for the homeless and friendless. As the public institutions became increasingly overcrowded, underfunded, and generally wretched places, middle-class families tried their best to avoid them, scrimping to pay for a private institution or managing to keep their family members at home.[50]

Randolph Bourne, author of the 1911 essay "The Handicapped: By One of Them," became a prominent public figure as a writer and critic at this unpropitious time. Barely five feet tall, Bourne had a facial disfigurement from a difficult birth and curvature of the spine from childhood spinal tuberculosis. Though he had little functional impairment, as Paul Longmore has written, "people reacted primarily, and often with extreme aversion, to his appearance." Ellery Sedgwick, his editor at the *Atlantic Monthly*, "could not overcome his revulsion and invite the young man to stay for luncheon at New York's exclusive Century Club." It was the same everywhere he went: "in Paris, concierges catching sight of him repeatedly refused him lodgings, until after two days he finally found a vermin-ridden flat. At Columbia University, recalled a friend, some were 'instinctively hostile to him, either because of his radical ideas, or because of his personal appearance.'"[51] The poet Amy Lowell spoke out loud the usually tacit connection understood to exist between the body and the inner man: "his writing shows he is a cripple."[52]

Making the Invisible Visible

It is an old and intuitive idea that the hidden inner self is manifested in external appearance, but its significance and connotations are adapted to suit the needs of particular times and places. For eugenicists, beauty became an important marker for superior genetic inheritance, associated with keener intelligence, superior morality, and greater energy. Ugliness, conversely, was a signal of eugenic danger. Thus, when public figures debated immigration policy or officials evaluated the desirability of certain types of immigrants, appearance figured prominently. It was not simply that outer defects inspired repugnance, but that they might be signs of inner corruption. Germ theory at the time was bringing to public consciousness the danger posed by the microscopic world. Cholera, tuberculosis, smallpox, diphtheria, influenza, polio, and other infectious diseases were dreaded everywhere, but especially in the crowded cities. Now the gene—or "germ plasm," as it was usually called—threatened not just illness in the present but an ever-widening stream of polluted heredity poisoning the future. This invisible enemy seemed, if anything, even more insidious and difficult to combat. It was just as crucial to quarantine the infectious gene as it was the infectious germ. And neither could be seen except by its symptoms.

In his bestselling 1924 book on eugenics, *Fruit of the Family Tree*, Albert Wiggam explained that "beauty and intelligence are probably linked together in the very inner processes of the evolution of organic life," that "good-looking people are better morally, on the average, than ugly people," and that "every increase of beauty will mean an increase of bodily and mental energy." Outer appearance was the single best indicator of inner soundness, and if "men and women should select mates solely for beauty, it would increase all the other good qualities of the race." Wiggam found plenty of reason, then, to worry about immigration, for the "nobility of any civilization can, to a considerable extent, be measured by the beauty of its women and the physical perfection of its men," and although "ugly women in America" had once been rare, America was now "getting them in millions." He urged readers to "examine these women as they are unloaded at Ellis Island" (a phrase that made them sound something like livestock): "scarcely one in hundreds would be called beautiful. They are broad-hipped, stout-legged with big feet; broad-backed, flatchested with necks like a prize fighter and with faces expressionless and devoid of beauty." To the notion that "beauty is only skin deep," Wiggam countered that "beauty is as deep as the human soul, as deep as evolution itself."[53]

For advocates of restriction, the subject of the appearance of immigrants was often couched in terms of visual disgust. Edward Alsworth Ross, a prominent professor of sociology at the University of Wisconsin, observed, "To one accustomed to the aspect of the normal American population, the Caliban type shows up with a frequency that is startling." Their defective inner nature was as plain as their faces, for they were "hirsute, low-browed, big-faced persons of obviously low mentality." These conclusions, he insisted, were not based on observing new arrivals when they were disheveled and worn out from the long journey, but later at social gatherings, "washed, combed, and in their Sunday best." Good grooming, however, did not conceal from him their true nature, for to "the practised eye, the physiognomy of certain groups unmistakably proclaims inferiority of type." Ross was also struck that among the women, "beauty, aside from the fleeting, epidermal bloom of girlhood, was quite lacking." Facial defects abounded: "in every face there was something wrong—lips thick, mouth coarse, upper lip too long, cheek-bones too high, chin poorly formed, the bridge of the nose hollowed, the base of the nose tilted, or else the whole face prognathous. There were so many sugar-loaf heads, moon-faces, slit mouths, lantern-jaws, and goose-bill noses that one might imagine a malicious jinn had amused himself by casting human beings in a set of skew-molds discarded by the Creator." Were immigration to continue unabated, the inevitable result would be a "falling off in the frequency of good looks" in America.

Ross went out of his way to emphasize that it was not a question of ugly races: "one ought to see the horror on the face of a fine-looking Italian or Hungarian consul when one asks him innocently, 'Is the physiognomy of these immigrants typical of your people?'" Degenerate types were found in every race, but in the past few of them found their way to America. Now that was changing, thanks to the comparative ease and reduced cost of travel and the lure of unskilled jobs in the cities. The most conspicuous result was a lowering of the standard of beauty: "it is unthinkable that so many persons with crooked faces, coarse mouths, bad noses, heavy jaws, and low foreheads can mingle their heredity with ours without making personal beauty yet more rare. So much ugliness is at last bound to work to the surface."[54]

Immigration officials expressed similar ideas. Alfred Reed, a Public Health Service physician at Ellis Island, thought that "no one can stand at Ellis Island and see the physical and mental wrecks who are stopped there . . . without becoming a firm believer in restriction."[55] The role of appearances took many forms, however, not just in physical beauty. For immigration officials charged with picking out the undesirable types, social class was the first indicator of the likelihood of fitness or defect. First-class passengers enjoyed a presumption of fitness. A 1906 Immigration Bureau report noted that first-class "cabin inspection is limited except in certain instances to circulating among them, and thus observing them." Still, traveling first class did not assure unimpeded entry. As immigration officials pointed out, "If a passenger is seen in the first cabin, but his appearance stamps him as belonging in the steerage or second cabin, his examination usually follows."[56] For those whose appearance matched their cabin status, obvious signs of disease or defect would also merit a closer look, but inspectors moved with care: offending a well-heeled traveler risked adverse publicity and inquiries from politicians, journalists, and friends in high places.

In 1906 an elderly British woman, Lydia Mary Thompson, sailing first class on her way to Manhattan to visit family, was certified for insanity at Ellis Island. She apparently had been much frightened by a rough and stormy passage and had found it difficult to sleep owing to the pitching of the ship and noise from the saloon nearby her quarters. She was prescribed a sleeping draught by the ship's doctor to which she reacted badly and arrived in New York in a state of delirium and exhaustion. Noting her condition, a public health physician brought her ashore to be more closely examined, where she was certified as insane and refused entry. Thompson, however, had well-connected relatives in New York who summoned Charles Dana, one of the most prominent neurologists in the country and a professor of nervous diseases at Cornell University, to offer a second opinion. He found her "perfectly sane" and persuaded immigration officials both to reconsider and to allow her to await a final decision at the home of her relatives. This was, of course, highly unusual, but Commissioner Watchorn helpfully clarified the situation by telling the *New York Times* that she was "in the care of her friends, but constructively she is still on the ship."[57]

Frederick Peterson, a professor of neurology and insanity at Columbia University, next came to look in on Thompson and likewise found her to be "sane and mentally competent." A third professor and physician then paid her a visit, after which he wrote a five-page letter on her behalf that began:

> I, M[oses] Allen Starr, M.D., LL.D., Professor of diseases of the mind and nervous system in the College of Physicians and Surgeons, the medical Department of Columbia University in the City of New York; consulting physician to the Presbyterian, St. Vincent's, and St. Mary's hospitals, and the N.Y. Eye and Ear Infirmary; specialist in mental diseases; and corresponding member of the London Neurological Society and of the Paris Society of Neurologists, and formerly president of the American Neurological Association, and of the New York Neurological Society, declare that I have to-day examined Mrs. James Thompson of Somersetshire, England; and that I find her of perfectly sound mind in every particular.[58]

The illustrious neurologist expressed his dismay that "although she was a first-class passenger and accustomed to comforts," Thompson had been "made to wait in a crowded, hot, ill ventilated room in the midst of emigrants" and then had been "asked a number of confusing questions which she regarded as insulting, as they were about her private affairs." Having upbraided the officials for treating her like any common immigrant, Starr pronounced that "any procedure for her deportation as an alien insane person would be wholly unjustifiable." Three public health officials subsequently made the trip to Manhattan to reexamine Thompson in the comfort of her relatives' home, where they found her much improved. While the officials defended the original certificate as accurately reflecting her condition at the time of her landing, they unanimously recommended that she be allowed to stay.[59]

Like those in first class, second-class passengers were looked over on board ship and subjected to a less thorough examination "owing to the lack of accommodations, of light, of space," which, one inspector noted, could not but "detract from the efficiency of the examination."[60] Inspectors did their best under the circumstances, knowing that ineligible immigrants sometimes came second class hoping to improve their chances. The vast majority, however, traveled steerage and upon arrival went through what was known as "line inspection." Inspectors, who prided themselves on their ability to make a "snapshot diagnosis," had only a few seconds to spot the signs of disability or disease as immigrants streamed past them single file. The instructions issued for the guidance of inspectors indicated that, if possible, "each individual should be seen first at rest and then in motion" in order to detect "irregularities in movement" and "abnormalities of any description."[61] An Ellis Island inspector explained that "the gait and general appearance suggest health or disease to the practised eye, and aliens who do not appear normal are turned aside, with those who are palpably defective, and more thoroughly examined later."[62] Inspectors tried to observe immigrants as they carried their luggage to see if "the exertion would reveal deformities and defective posture."[63]

Eugene H. Mullan, an Ellis Island medical officer, wrote that it was his job in those few seconds "to look for all defects, both mental and physical, in the passing immigrant. As the immigrant approaches the officer gives him a quick glance. Experience enables him in that one glance to take in six details, namely, the scalp, face, neck, hands, gait, and general condition, both mental and physical." If anything at all "abnormal" caught his notice, he pulled the immigrant aside; for example, "immigrants that are thin and of incertain [*sic*] physical make-up are stopped while the officer comes to a conclusion as to the advisability of detaining them for further physical examination."[64] Anything visually out of the ordinary increased the probability that an immigrant would be taken aside and closely examined, which then made it more likely that some problem might be discovered. Inspector S. B. Grubbs claimed that line inspection often hinged on extraordinary intuition and that many times "the keenest of these medical detectives did not know just why they suspected at a glance a handicap which later might require a week to prove."[65] Inspector Victor Safford maintained that it was "no more difficult to detect poorly built, defective or broken down human beings than to recognize a cheap or defective automobile"; indeed, the skilled examiner could often identify defects from "twenty-five feet away" by observing a "man's posture, a movement of his head or the appearance of his ears, requiring only a fraction of a second."[66]

Journalists such as Frederick Haskin were much taken with the story of canny inspectors exercising their skills of detection on the line:

> As they approach, the doctors begin to size up each immigrant. First they survey him as a whole. If the general impression is favourable they cast their eyes at his feet, to see if they are all right. Then come his legs, his body, his hands, his arms, his face, his eyes, and his head. While the immigrant has been walking the twenty feet the doctors have asked and answered in their own mind several hundred questions. If the immigrant reveals any intimation of disease, if he has any deformity, even down to a crooked finger, the fact is noticed. If he is so evidently a healthy person that the examination reveals no reason why he should be held, he is passed on. But if there is the least suspicion in the minds of the doctors that there is anything at all wrong with him, a chalk mark is placed upon the lapel of his coat.[67]

The *Portland Oregonian* similarly lauded the "visual tests" skillfully applied by inspectors "experienced in the detection of outward signs of unfitness" and quoted an Ellis Island inspector who claimed that "the faces of the various types of psychopathic cases are as open books to the experienced examiner."[68] As much as officials evinced pride in their abilities, they occasionally admitted that the system was far from perfect. Most often they did so in the context of asking for greater resources. There were never enough inspectors for the large numbers passing under their gaze. Under questioning before a House of Representatives committee, the assistant surgeon general of the Public Health Service conceded that the typical inspection was "cursory" and "mainly directed toward detection of the obvious physical defects, such as the lame, the blind, the deaf, or for the purpose of detecting mental defects."[69]

Nothing in the law explicitly stated that feelings of attraction or repulsion toward immigrants' appearance should carry any weight, yet inevitably they mattered a good deal. Given what is now known about the influence of appearance on the attitudes of judges and jurors toward defendants, bank officers toward loan applicants, teachers and professors toward students, even parents toward their own children, it is hard to imagine that appearance would not have influenced immigration officials.[70] This would be especially true at the initial stage of picking out immigrants for closer inspection on the basis of a glance and intuition, but appearance mattered throughout the entire process, even in the files that were sent to Washington, D.C., to accompany appeals. Robert Watchorn, recommending exclusion of Chune Fruman, noted that he was a "man of very poor physical appearance." William Williams recorded that the curved spine of Bernhard A. Mydland was "a physical handicap," but it was in his favor that it did "not show up as much under his clothing as when naked."[71] Subjective judgments of immigrants' appearance were considered relevant and significant information, and comments such as "his appearance is distinctly against him," "appearance very poor," "appearance is not good," and "presents a very poor physical appearance" were routinely included.[72] None of this was considered invidious or prejudicial, for even if inspectors did not discriminate on the basis of appearance, employers would: "if one man comes in healthy, and another has a deformity or a horrible disease on his face, the latter would not stand a chance of getting work. They would accept the good man. Does not that affect his ability to get employment?" the surgeon general asked rhetorically in 1907 at a meeting of immigration officials.[73]

In some cases, the reason for exclusion appears to have been almost entirely based on appearance. In March 1905 Domenico Rocco Vozzo, a thirty-five-year-old Italian immigrant, was puzzled to find himself barred from entering the country at the port of Boston. Vozzo was a "bird of passage," a migrant worker who intended to earn some money and then return to Italy. It was his second trip to the United States, and he had encountered no difficulty his first time three years earlier. But the medical inspector certified him for "debility," that is, "lack of force and vigor," and he was excluded as likely to become a public charge. Vozzo retained an attorney who maintained in his letter of appeal that Vozzo was strong, robust, and healthy. In fact, he "looks perfectly healthy below the head," but has a "curiously shaped head, and his skin looks rather white, almost bleached, and his ears are quite thin." He had never been ill, had always worked, and during his recent two-year sojourn in the United States had fully supported himself while saving money. He brought with him $20 in cash and had friends who had filed affidavits on his behalf. The commissioner at the Boston station, however, recommended against admission and sent to the secretary this evidence: "I enclose his picture which I think will convince you that he is not a desirable acquisition." Vozzo was sent back to Italy. (The photograph, still in the file, is a bit faded now, but aside from several days' beard growth and a scowl, it is hard to see what the commissioner was referring to.)[74]

In February 1913 Felix Petkos was refused entry at the port of Boston. The examining physician had merely stated on his certificate that Petkos had "psoriasis, a chronic noncommunicable skin disease, which will cause him to seek treatment from time to time." That was all—nothing about its affecting his ability to earn a living. At his first hearing, the board concluded that inasmuch as it is "a well-known fact among laymen that the odor of this particular disease is so annoying . . . it would be very difficult for him to get employment," and ordered Petkos excluded "first, as likely to become a public charge, and second, as afflicted with a physical defect which will affect his ability to earn a living." At a rehearing two days later, the board affirmed the decision to exclude "for previous reasons except it is the opinion the appearance of the disease rather than the odor will be offensive." After the Secretary of Commerce and Labor rejected Petkos's appeal, his lawyer filed a writ of habeas corpus with the Massachusetts District Court. After hearing testimony from medical experts, the court concluded that the "authorities in fact had no actual knowledge of the character of the disease," that it is "attended with no odor," that usually it does "not appear on exposed part of the body," and that it generally "does not interfere with the sufferer's ability to labor." Exclusion on the basis of appearance was acceptable, the court noted, but in this case the examiners were simply mistaken about the effect on appearance of this particular disease. The government appealed the case, however, to the Circuit Court of Appeals, which found that questions of fact are to be determined by immigration officials, not the courts, and reversed the district court's decision.[75]

The principle that persons of unsatisfactory appearance were undesirable was, of course, impossible to apply with any consistency. In 1906, for example, Abram Hoffman, a twenty-five-year-old tailor from Russia, was ruled likely to become a public charge by the immigration board because of his curved spine. His attorney labeled this notion "ridiculous and absurd." "The axiom," he continued, "that one who is unfortunate enough to suffer from a certain infirmity, is likely for that reason alone, to become a public charge, is entirely new to us." Warming to the subject, the attorney asked, "Are we living in this enlightened Twentieth Century where everyone is supposed to be given a fair opportunity, or are we going back to the times of the Salem witch-craft, when, because a woman was old and afflicted with a high back, she was considered and treated as a witch?" He added that the "immigrant's affliction can in no wise affect his earning capacity as a tailor." Commissioner Watchorn was ambivalent, in part because the evidence based on appearance was conflicting. On the one hand, visually "the spinal curvature for which he is certified is quite obvious." On the other, Hoffman was "well dressed, and came as a second cabin passenger." With appearance working both against and in favor of the immigrant, Watchorn made the unusual decision to forward the case files to Washington without recommendation. In his summary for the secretary, the commissioner general reiterated that Hoffman made a good appearance, and that the "Commissioner states that the alien is intelligent looking and is well dressed; he came in second cabin." In this case, the positive aspects of his appearance and social class trumped the negative appearance of his disability. Hoffman was admitted on appeal.[76]

Otto Engel was not so fortunate. Arriving from Germany in 1905, just five months before Hoffman, he too had pronounced curvature of the spine. Engel was a printer, which meant that he would likely enjoy higher pay and status than a tailor such as Hoffman. Commissioner Watchorn had interviewed Engel and was "very favorably impressed with the man's appearance," according to Assistant Commissioner Murray, who added that Engel was "apparently strong, intelligent and clean." He recommended admission. Nevertheless, the same commissioner general and Secretary of Commerce and Labor who admitted Hoffman rejected Engel's appeal. Nothing in the record clearly explains the different outcomes. It is unlikely that it had anything to do with nationality or race, for German Jews such as Engel usually were, as a general matter, held in higher regard than Russian Jews such as Hoffman. Engel had a sister and a brother-in-law willing to help if needed, and Hoffman had a brother and a sister-in-law willing to do so. The one significant difference between the two is that an attorney for Hoffman (likely provided by the Hebrew Immigrant Aid Society) submitted a typed three-page "Brief for Appellant." It was eloquent and stirring but largely irrelevant, other than the information that Hoffman had relatives willing to help him. Engel handwrote a half-page appeal himself, containing the same quantity of relevant content. There is no way of knowing if that made the difference, but appearances can take many forms.[77]

Mental Defect

As with physical disabilities, inspectors scanned arriving immigrants for telltale visual signs of insanity and feeblemindedness. Medical inspectors were ill-prepared to diagnose either. The "Book of Instructions for Medical Inspection" published in 1910 merely noted that "medical officers are referred to appropriate text-books for details of the methods of examination for mental disease." Still, staff physicians generally expressed confidence in their ability to spot visual clues. One medical officer in 1905 explained that "if the manner seems unduly animated, apathetic, supercilious or apprehensive," the immigrant was taken aside for closer examination. "A tremor of the lips . . . , a hint of negativism or retardation, an oddity of dress, or an unusual decoration worn on the clothing—any is sufficient to arouse suspicion." Looking back twenty years later, however, he was less confident: "I had a little knowledge of psychiatry in my head, a little piece of chalk in my hand, and four seconds of time. I was supposed to take that little knowledge, that little piece of chalk and little time, and mark with the chalk an 'x' on the shoulder of every applicant I thought should be held for further mental examination."[78] The procedure for "mental examination was always haphazard," another official later recalled in an interview with the historian Alan Kraut; "it couldn't be any other way because of the time given to pass the immigrants along the line. Some questioning—if the immigrant did not respond or looked abnormal he was sent in and given a further examination."[79]

Immigration officials and Public Health Service physicians were under constant pressure to improve their ability to detect and exclude the feebleminded. The State Charities Aid Association of New York issued a resolution in 1912 calling for "mental examination of arriving immigrants by physicians trained in the diagnosis of insanity and mental defect," and for "adequate facilities for the detention and examination of immigrants in whom insanity or mental defect are suspected."[80] That same year a conference of alienists (psychiatrists) and social workers issued a recommendation to Congress to require that Public Health Service officers conduct onboard examinations "during the voyage with special reference to their mental condition," that all ports be provided with staff who have had "special training and experience in the detection of insanity and mental defectiveness," and to extend the period for expulsion of landed immigrants from three years to five.[81] An editorial in the *Journal of the American Medical Association* in 1913 complained that too many mental defectives were coming into the country and filling the schools and institutions, and that physicians at Ellis Island made only "the most cursory sort of inspection," which made it "highly probable that large numbers of the higher grades of mental defect must pass unobserved."[82] The Massachusetts Medical Society and the Medical Society of New York both wrote to the Immigration Committee of the House of Representatives in 1914 to protest the number of "mentally defective" immigrants making it past examiners only to "propagate and so deteriorate the mental health of the Nation."[83]

Although immigration officials defended their system of visual detection, they too worried that numerous defectives might be going undetected. In 1905, when three immigrants became public charges within three months of arrival and were diagnosed as insane, Commissioner-General Sargent took the Ellis Island commissioner to task. "It is passing strange," Sargent wrote to the commissioner, "that these persons were admitted as being in good physical and mental health, when they have been certified as suffering from insanity . . . within such a comparatively brief time." He demanded an explanation "for this condition of affairs." The chief medical officer responded that, regardless of how many immigrants are certified as insane, "it is quite probable that as many more have passed," and that this was inevitable as long as inspectors had to decide who to pull aside for closer inspection "almost as rapidly as he can count the immigrants coming along the line."[84] As much as inspectors prided themselves on their snapshot diagnoses, when criticized for missing telltale signs, they emphasized the difficulty of the task.

By the 1910s intelligence testing was coming into its own with the development of the Binet-Simon intelligence tests, which the immigration service tried to adapt in ways that would compensate for cultural bias. The problem was that they could not possibly test a significant number of individuals without bringing the entire machinery of inspection to a standstill. For the initial inspection, officials had no choice but to rely on their own idiosyncratic and intuitive methods of spotting signs of deficiency or deviance. In his annual reports of 1910 and 1911, William Williams complained of the "lack of time and facilities for thorough examination," stressing the myriad social problems that could be caused by a feebleminded immigrant getting past his physicians, from dependence on public support to criminality, pyromania, and propagation of "a vicious strain that will lead to misery and loss in future generations."[85] Carlisle Knight, a medical officer, explained in 1913 that the "idiot is a type which is usually easily recognized" and the "imbecile is a step further up the scale from the idiot [and] the stigmata of degeneration in these cases are generally well marked," but that detecting "that class of feebleminded commonly known as the 'school dunce'" was another matter. They simply did not stand out visually, and there was not enough time to question, let alone test, everyone.[86]

Inspector Mullan's technique was to approach "inattentive and stupid-looking aliens" and question them about their age, destination, and nationality, in some cases following up with simple questions in addition and multiplication. If anyone should "appear stupid and inattentive to such an extent that mental defect is suspected, an X is made with chalk on his coat" to indicate further examination was warranted. Mullan, like many other inspectors, argued that once they arrived at the stage of actually questioning immigrants individually, experienced medical officers were able to reach reliable conclusions, even calibrating their evaluation according to the race of the immigrant. It sometimes happened that they approached one whose "facial expression and manner [was] peculiar" enough to indicate mental illness, only to discover "that the alien in question belongs to an entirely different race" from the one they had assumed. At that point they would realize that the "peculiar attitude of the alien in question is no longer peculiar; it is readily accounted for by racial considerations."[87]

Not all agreed, however, that officials were as adept as all that. Fiorella La Guardia, who in 1934 became mayor of New York City, worked as an interpreter at Ellis Island from 1907 to 1910; he later wrote in his memoirs, "I felt then, and I feel the same today, that over fifty percent of the deportations for alleged mental disease were unjustified. Many of those classified as mental cases were so classified because of ignorance on the part of the immigrants or the doctors and the inability of the doctors to understand the particular immigrant's norm or standard."[88] Immigration officials themselves sometimes conceded that reading intelligence and mental soundness in the face and body was unreliable, but they saw it as the only practicable means of screening large numbers of people. In 1911 Ellis Island officials asked the superintendent of the New Jersey Training School at Vineland, Edward Johnstone, and his research director, Henry Goddard, whether they could offer any advice on ways of better detecting mental defectives. They could not, they responded, but the following year Goddard carried out an experiment, assigning one of his Vineland assistants to pick out arriving immigrants who appeared defective, while Ellis Island officials did the same. Another assistant followed up by administering the Binet-Simon test to all those selected. Goddard found that his team was successful 80 percent of the time, while the Public Health Service physicians achieved less than 50 percent accuracy.[89]

Knox, a medical officer at Ellis Island from 1912 to 1916 and a self-taught expert in the visual detection of mental defect, was skeptical of Goddard's results, believing the Binet test to be culturally biased, and set out to develop his own set of tests. (Goddard inadvertently proved Knox correct the following year when he found, using the Binet-Simon test, that approximately four out of five Jewish, Russian, Italian, and Hungarian immigrants were feebleminded.) By 1913 the Public Health Service had decided that a new manual of procedures for mental examination was needed. To circumvent the cultural biases of the Binet-Simon test, Knox developed a series of intelligence tests that used significant nonverbal components involving puzzles, wooden blocks, and simple arithmetic problems.[90] Nonetheless, to be practical, the tests could only be given to those who had first been picked out visually, and for this purpose he published a "Diagnostic Study of the Face" as a guide to the "facial signs" of mental defect.[91]

Fig. 4.—Low moron. Age, 30 years.

Fig. 8.—Dementia precox. Apathy.

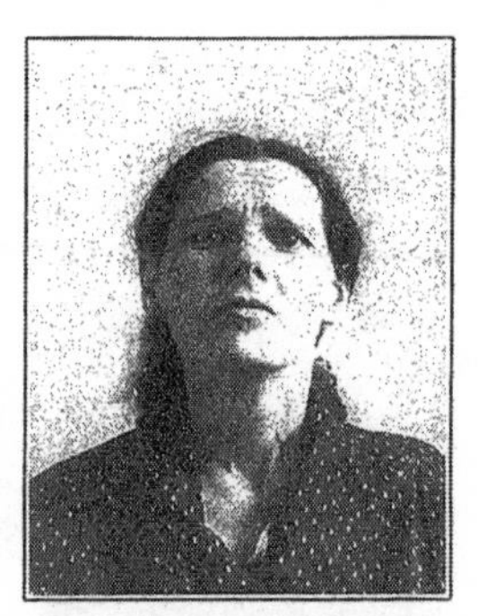

Fig. 10.—A typical expression of anxiety.

Fig. 5.—A constitutional inferior.

'ig 9.—Irritability and surliness. Th firm mouth and earnestness of expres sion suggest combativeness.

Fig. 11.—Advanced juvenile paretic. Expression of apprehension. The facial expression suggested feeble-mindedness.

FIGURE 8. This Public Health Service manual urged inspectors to practice making "close observations of facial expressions, both in normal and abnormal persons" in order to develop their ability to spot mentally defective immigrants. It provided photographs purporting to illustrate various conditions. United States Public Health Service, *Manual of the Mental Examination of Aliens* (Washington, DC: Government Printing Office, 1918).

Sexual Disgust

Tests might measure intelligence, inspection ferret out physical defects, and facial expressions betray an unbalanced mind to the practiced eye, but an immoral character lay hidden and resisted discovery. One of the attributes of the immoral person, after all, was a heightened capacity to dissemble and present a false face. Finding means to divine the inner moral sense from the outer appearance was crucial, especially in cases of sexual perversity. Although Knox claimed an ability to deduce from certain "facial indications" the existence of "unusual sexual proclivities in young males" (and supposedly confirmed his ability to read those signs in one case when, upon closer examination of a suspected "pervert," he discovered "abdominal tattooing of a sensual nature"), inspectors generally had little to go on other than vague stereotypes.[92]

Donabet Mousekian, an Armenian Turk, stood before the Board of Special Inquiry on April 23, 1905, having been diagnosed with "feminism" (also referred to as "lack of sexual development" and "arrested sexual development"). In Mousekian's case this meant an absence of male sexual organs, although in others it referred simply to subnormal development. Mousekian's hearing was extraordinarily brief. No one mentioned the diagnosis or questioned him about it. An embarrassed silence seemed to fall over members of the board. After asking only the most basic questions concerning his identity and background and noting that he brought $48 with him, the transcript of the hearing reads as follows:

INSPECTOR 1: In view of the Doctor's certificate I move to exclude him as likely to become a public charge.
INSPECTOR 2: Second motion.
INSPECTOR 3: Excluded.

This was all the hearing that Mousekian received.

In his appeal, Mousekian explained that he had fled the violent oppression of Armenians in Turkey and had officially renounced his citizenship. If he were to return, he would be immediately imprisoned. He wrote, "It would be much better that you kill me." All of his relatives had already come to America, including his two brothers, with whom he was going to live. One was a citizen, the other was scheduled to obtain citizenship the next month, and both were well employed. Mousekian was a photographer by trade, as well as a weaver and dyer of rugs, and an experienced restaurant cook, and could work at any of these trades. He wrote, "I am not ill, have no contagious disease; my eyes, feet, hands and ears are sound; only I am deprived of male organs; this is not a fault because it has come from God and my mother: what can I do? It won't do any harm to my working . . . what harm can I do to the U.S. by my being deprived of male organs?" His brothers wrote letters in much the same vein, asking plaintively, "How is it his fault? Our father and mother are dead; he is our only brother . . . ; we guarantee that he will not be a public charge; we are able to give the required guarantee; he can not return to Turkey; we are US citizens, hence we beg US government not to separate our brother from us." Their attorney telegraphed the bureau that "they have sufficient means to provide bond for maintenance or purposes of appeal."

Commissioner Watchorn wrote in his letter to accompany the appeal that "appellant is devoid of every external evidence of desirability" and is "really repulsive in appearance." His sense of disgust with Mousekian was further conveyed by his unwillingness to name Mousekian's condition. Instead, he confined himself to simply stating that the medical certificate furnished "sufficient indication of his physical defects." No one but the physician who had to write out the certificate, and Mousekian himself, ever said a word about the nature of his defects. Mousekian was returned to Turkey where, if he lived that long, he would have been caught up in the Armenian holocaust ten years later.[93]

Varied justifications for exclusion were offered up in these kinds of cases. Mousekian was excluded by the board as likely to become a public charge, which was the default justification for exclusion on the basis of physical disability. Watchorn's letter, however, also referred to section 2 of the Act of 1903 as "beyond a doubt having precluded the possibility of his lawful admission." There are three parts of this section that could plausibly be applied to Mousekian. One was the public-charge provision, which would seem the most likely except for Watchorn's comment that the law "beyond a doubt" precluded admission. No evidence was presented that Mousekian was clearly unable to work, and no immigrant with his skills and family support, but without his defect, would have been excluded on those grounds. The second possibility is the prohibition of "persons afflicted with a loathsome or with a dangerous contagious disease." No doubt Mousekian was seen as "loathsome" by the officials involved, but his condition is described in immigration documents as a "defect," not a "disease." A third possibility is that Watchorn could have been referring to the mandatory exclusion of "insane persons." This might seem implausible, but the cases of Arotioun Caracachian in 1908 and Spagliono Francesco in 1911 bring it within the realm of possibility. Caracachian, an Armenian Turk, was certified for "arrested sexual development." When the board asked the inspecting physician whether he had found any "marked degree of degeneracy," he replied unequivocally that the "alien is a physical degenerate," adding that although he had not examined him for "mental degeneracy . . . generally speaking they do not come up to the normal standing."[94] Francesco was excludable on multiple grounds. He was a fifteen-year-old traveling alone, in itself sufficient for exclusion. Moreover, he had never worked or attended school, could not read or write, and, in addition to having "arrested sexual development" and "a very poor appearance," was described as "undersized [and] frail." His exclusion was a foregone conclusion. But the physician on duty chose to elaborate on the import of his condition: "persons so affected are liable, owing to inability to satisfactorily perform sexual congress, to become addicted to unnatural practices in this respect, with accompanying mental deterioration, which in some instances may lead to actual insanity."[95] The rejections of Mousekian, Caracachian, and Francesco were most likely based on the belief that they would become public charges as well as repulsion at their condition and appearance, but it is possible that the decisions were based in part on the long-popular idea of masturbation-induced insanity.

Predisposition to insanity was invoked again when Nicolaos Xilomenos, a nineteen-year-old Greek immigrant, was certified for "lack of sexual development" in 1912. The board concluded that although "he arrives with a fair amount of money," he would be unlikely to find employment due to his "frail physical appearance and certified condition." Two of his cousins, owners of small lunch counters, testified that they would help him, one of them filing an affidavit promising board and lodging, assistance in finding work, and a bond to guarantee that he would not become a public charge. The following day, this same cousin obtained a written promise of employment from a fellow restaurant owner and an offer to provide an affidavit swearing to it. Commissioner Williams, rather than addressing the particulars of the case, instead sent a statement from one of his medical officers that "may be said to apply in general to all cases thus certified" to the effect that those so "afflicted are occasionally sexual perverts," may suffer "melancholia" or "mental exaltation," and "are not infrequently mental [*sic*] unstable." Although they may "appear strong and robust," that does not reduce "the liability of perversions or mental instability." Xilomenos was turned back. Whether it was due to the likelihood of mental instability, sexual perversion, or becoming a public charge was never made clear.[96]

The reluctance of officials to openly discuss this diagnosis worked to the advantage of Israel Raskin when he was refused entry in 1922 for "lack of sexual development which may affect his ability to earn a living." Nothing was said about it at his hearing. The commissioner general's response to his appeal noted simply that the Public Health Service was "decidedly opposed to the admission of aliens suffering from the defect certified in this case, as it is usually a forerunner of insanity." And so, when Raskin's sister, Rose, an American citizen living in Chicago, petitioned the New York District Court on his behalf for a writ of habeas corpus , she was able to make three powerful arguments for his release: that the "matters relied upon to cause the judgment of expulsion and forcible deportation were not specifically referred to" at the hearing; that the decision had been "based on an undisclosed assumption"; and that Raskin had not been informed of the reason for his exclusion and therefore was given no opportunity to dispute it. The judge found this reasoning persuasive and ordered Raskin released.[97]

In preparation for a possible appeal of the case, the commissioner general asked the surgeon general to clarify for him, in a memorandum, the reasons people like Raskin were excluded. His response briefly mentioned that they are "predisposed to abnormal sexual conduct, and are usually mentally below par," but otherwise played down the unsubstantiated connections to insanity that immigration officials had drawn. Instead, he emphasized the "likely to become a public charge" justification: "these individuals are rather frequently encountered at Ellis Island, and it has been, so far as I am aware, the unanimous testimony of the expert alienists stationed there, that these persons present bad economic risks." The risk arose not because they were disabled from work in any concrete way, but rather because "their abnormality soon becomes known to their associates who make them the butt of coarse jokes to their own despair, and to the impairment of the work in hand. Since this is recognized pretty generally among employers, it is difficult for these unfortunates to get or retain jobs, their facial and bodily appearance, at least in adult life, furnishing a patent advertisement of their condition."[98] It was not a question of their ability to work. The issue was that they might they face mockery and discrimination due to their abnormal appearance and thereby be pushed into poverty. Their disability originated in social relations, but the surgeon general deftly relocated it to their anomalous bodies.

Other forms of ambiguous sexuality were, like "feminism," treated with ill-concealed distaste, along with suspicion that they signified an immoral nature. In 1908, at the port of Philadelphia, Helena Bartnikowska was held for close examination, after which the examining physician testified that "this supposed woman . . . is found to be suffering from Hermaphrodism," adding significantly that her voice was masculine and that she had facial hair. "Such cases," he continued, "are usually of perverted sexual instincts, and with lack of moral responsibility." She was a young woman of twenty-two years and identified as "Russian Polish," although between the ages of one and nine she had lived with her parents in Cleveland until the family returned to Europe. Now her aunt and uncle had invited her to come back to Cleveland to live with them and had paid for her passage. A board member asked her, "Is your aunt aware that you are the kind of person you are?" Bartnikowska answered, "She does know. My aunt knows." The board pressed her: "As a matter of fact, were you deported from the United States; did the Government send you back to the country whence you came?" "No sir," she answered. The next question was, "Were you ever in an institution of any kind?" Her response suggests that she did not understand the drift or import of the question: "I have never been in any institution or in any hospital; never had any trouble with my eyes or anything else." The inspector obviously was not asking about her "eyes or anything else" having to do with her health, but rather whether she had been institutionalized for the defect that seemed, in his eyes, to define her. Moving to the question of heredity, she was asked whether her mother or father had ever been in institutions. Neither had been, and about her mother she added that, a few years earlier, "she just took sick and died," again seeming to miss the implication of the question. The board voted unanimously that, in view of the "Doctor's certificate and statement, especially the portion which deals with the moral responsibility of the alien," Bartnikowska should be "excluded as likely to become a public charge."

On the medical certificate filled out by the examining physician, however, after the standard printed text, "This is to certify that the above-described immigrant has___," the physician had written his diagnosis, "hermaphrodism," and then crossed out the printed text that followed, "affecting ability to earn a living." That is, he specifically ruled out any impact on her ability to work and in his testimony emphasized only the probability of sexual perversion. Furthermore, her uncle and aunt had submitted an affidavit to affirm their citizenship and sound financial standing and to pledge that their niece "shall be given proper care, clothing, board, [and] medical attention and shall never be permitted to become a public charge." Yet she was excluded as likely to become a public charge. The uncle and aunt were perplexed, and in their letter of appeal asked, "Whereas the complaint made does not cover a loathsome or contagious disease and whereas her maintenance is assured," why has she been refused admittance? In his own letter to Washington, the Philadelphia commissioner simply quoted the physician's statement concerning sexual perversion, adding no other justification for his recommendation that the board's decision be upheld. The commissioner general handwrote at the bottom of this letter, "I think the appeal of this alien should be dismissed because of the Doctor's statement before the Board"; in his own letter, he concluded that in "view of this statement, the ability and willingness of the alleged relatives to support and maintain this alien notwithstanding, I recommend that the appeal be dismissed." (Relatives who did not appear in person before the board were always "alleged," regardless of corroborating evidence; it does not imply any particular skepticism in this case.) There was no discussion of how her hermaphroditism might affect her ability to work, nor any questioning of the ability and willingness of her relatives to support her. Nevertheless, the Secretary of Commerce and Labor refused her admittance on grounds that she was likely to become a public charge. The apparent incongruity is not explained in the record, but perhaps it was due to a belief that the "perverted sexual instincts" of her abnormal body would lead her into unsanctioned sexual practices, which could land her in jail or an institution. But perhaps also it had to do with disgust, with the physician testifying that "this person's voice is of a masculine type, with considerable beard on the face." Was it more troubling when he warned of her wayward sexual instincts, or when he described, in front of Bartnikowska, the three men of the board, and anyone else who dropped by that day, the "well-developed penis, with glands [*sic*] and foreskin, without external testicles," of "this supposed woman?"[99]

Conclusion

The precise number of those turned back for physical and mental defects in these years is difficult to pin down, owing in part to the assumptions that linked disability and dependency. Until 1908 exclusions based on physical defects were combined with those of nondefectives in the category of "likely to become a public charge." This was always the largest category of exclusion, but the criteria applied were never straightforward or clear-cut. Lack of cash on hand, although taken into consideration, was not in itself a primary factor. A congressional commission reported in 1911 that "pauperism among newly admitted immigrants is relatively at a minimum, owing to the fact that the present immigration law provides for the admission only of the able-bodied, or dependents whose support by relatives is assured."[1] The commissioner general that same year explained that "likely to become a charge upon the public" was a category "nearly approximating and often merging into the question of physical, mental and moral fitness."[2] Both suggest that most people were placed in this category because they were considered defective.

After 1908 a rejected immigrant was counted in the category of "mental or physical defective" if deemed merely defective, but in the "public charge" category if both defective *and* a likely public charge. Taken together, exclusions in both categories usually ranged from 0.5 percent to 1.5 percent of all immigrants, except during the years of World War I, when they reached nearly 5 percent. Wartime conditions dramatically reduced immigration, which in turn allowed more careful inspection, proving the argument of the Public Health Service that enforcement could be much more efficient with a better ratio of inspectors to immigrants.[3]

These figures are all just the tip of a much larger iceberg, however, and seriously understate the impact of the laws. Those who arrived on American shores to be inspected had already been through several screenings. First, American immigration laws were widely advertised abroad—many potential immigrants would not risk the hardship and expense of the journey knowing they might be sent back. Secondly, not only were shipping companies required after 1891 to return rejected immigrants to their port of embarkation at no charge and to pay a fine for each, but the same applied if an immigrant later became a public charge or was otherwise discovered to have an excludable condition that initially passed unnoticed, up to a year after landing (increased to two years in 1903 and to three in 1907). After 1893 the law required the ship's surgeon to examine all passengers and the captain to certify that none appeared to be defective.[4] Finally, ticket agents in Europe also became de facto inspectors because, as the superintendent of immigration noted approvingly in 1894, shipping companies instructed agents to refuse tickets to "blind, lame, deaf and dumb, and crippled persons" and fined them for tickets sold to passengers subsequently rejected at boarding or upon arrival. There is good reason, then, to suppose that those turned away at the borders were a small fraction of those who would have immigrated to America but were deterred because of disability.[5]

Opinions varied on the efficacy of the inspections carried out by steamship companies, in part because inspections varied from company to company. Advocates of tighter controls over immigration always complained that the company inspections were inadequate, but then they had the same complaint about port inspections. No doubt they *were* imperfect, given the numbers of immigrants that shipping companies were dealing with, but according to most contemporary observers this initial screening process had a major impact. In 1893 Senator Henry Hansbrough maintained that "landings would have increased enormously but for the restrictive features of the law." Not only did "steamship agents report their refusal to sell tickets to 50,000 applicants" that year, but untold thousands more were "deterred from consulting ship agents" in the first place, knowing they would likely be rejected.[6] The commissioner general in 1894, Joseph Senner, believed that most shipping companies "earnestly endeavored to carry out the law, not only in the letter, but also in the spirit."[7] Prescott Hall found that the "chief value" of the law in 1897 was "in deterring those from coming to this country who might otherwise come."[8]

Preboarding inspection continued to improve after the turn of the century, according to immigration officials. William Williams reported early in 1904 that, as a result of increased fines for bringing defective immigrants, "far fewer undesirable immigrants are coming here now than came twelve months ago." In fact, rejections had decreased because "certain classes of people are now left behind who would otherwise necessarily be deported." Commissioner General Frank Sargent concurred that "passengers are being subjected to a more careful scrutiny on the part of the steamship companies before being taken on board, and a higher class of immigration is the consequence."[9] In 1906 Robert Watchorn reinforced the claim that the "refusal of steamship companies to carry undesirable immigrants is one of the greatest checks upon pernicious immigration."[10] In 1906 *Harper's Weekly* reported that, although it was "difficult to say just how many would-be emigrants were deterred from coming to the United States by the indirect operation of the American law at foreign ports," it was "safe to estimate . . . that there were several hundred thousand."[11] Finally, a federal commission in 1911 estimated that ten times as many were refused transportation for medical reasons as were barred at American ports, and that "the relatively small number of rejections at United States ports is good evidence of the effectiveness of the steamship-company inspections abroad." It attributed this to the fact that United States immigration laws were "well known among the emigrating classes of Europe," as was "the large number rejected at European ports, or refused admission after reaching the United States." That knowledge had "a decided influence in retarding emigration, and naturally that influence is most potent among those who doubt their ability to meet the law's requirements."[12] After all, the consequences for immigrants rejected after their journey were often dire. The *Chicago Inter Ocean* in 1900 described their plight:

> The final act of the deportation tragedy is played at the sailing of the return steamer. Many of the immigrants, before leaving home, sold everything they owned, converted it all into money for the passage. No homes are waiting for them. Their money is gone. Their hopes are dead. They are shut out from the new life of which they dreamed, and cannot even take up the old life where they left off with it. . . . In many ports the deported emigrants are dumped upon foreign soil, without money, and with no provision for reaching their old homes, which are, perhaps, across half of Europe from the port.[13]

The restrictive laws put in place before the 1920s prevented or deterred large numbers of Europeans from immigrating to the United States. Doubtless they kept out some immigrants who would have become public charges. Doubtless too they kept out others who would have been productive, self-supporting, contributing citizens. Where the balance point lay is impossible to tell. More important than numbers in this instance, however, is the history of the advocacy, passage, and enforcement of laws based on presuppositions born of fear and prejudice: emotional reactions to physical appearances and the hidden flaws they might reveal, beliefs about the social and economic value of disabled people and their potential for independent living, and an invidious equation of efficiency and speed with human worth and social welfare. The intentions of lawmakers—the ideas behind the laws and their administration, and the cultural context—are more to the point when the subject concerns questions of human rights. Amy Fairchild has persuasively argued that the inspection regime had a broader role than merely selection of qualified immigrants, that it educated both immigrants and the American public about the values of a competitive, capitalistic society: "It sorted immigrants publicly, baring them if need be, in an effective if not entirely conscious demonstration of the principles of efficiency, order, and cohesion that were expected of them."[14] In the same way, it educated everyone, immigrants and native-born Americans, disabled and nondisabled alike, about the worth of disabled persons. The multiple checkpoints for "sifting" and "winnowing" that it set in place reflected a similar and pervasive screening process that operated throughout the lives of many disabled people, including exclusion from education, community life, marriage, employment, and basic respect as citizens and human beings. The same kinds of legal, physical, and social barriers that prevented many disabled people from migrating freely also restricted their freedom and mobility within their own communities and societies.

After 1924 it becomes more difficult to trace the effects of immigration law on disabled immigrants. Beginning in that year, prospective immigrants were required to apply to local U.S. embassies or consulates for a visa and to submit to a full medical examination before they would be allowed to embark for the United States. Findings of excludability could not be appealed, and records of the individual examinations and interviews are not available, making it impossible to know how many or why disabled individuals were turned away.[15] Although inspection upon arrival continued as a second line of defense, the days of relying on a "snapshot diagnosis" were over, making it more difficult for immigrants to "pass" as nondisabled. Consular officials, according to Amy Fairchild, "sought to fill the quota for each country with only the most desirable immigrants," based on a principle of "selection with attention to 'quality.'" Officials were instructed to "pass only those who are found to be normal" and were given the latitude to refuse visas on the basis of any condition or defect that deviated from "normal" health and ability, without consideration of ability to earn a living. The Treasury Secretary concluded in 1925 that the "number of physical and mental disabilities and defects certified and the high percentage of refusal of visas for medical reasons" proved inspection abroad an effective means of "elimination of undesirable elements to our population." Between 1926 and 1930 nearly 5 percent of applicants were refused visas on medical grounds, several times higher than the typical rejection rate in earlier years, when the medical inspection took place solely upon arrival.[16]

Until 1990 immigration law still specified the exclusion of the "mentally retarded," "insane," and those with any "physical defect, disease, or disability" that "may affect" their ability to earn a living. In that year Congress removed most of the provisions that specifically targeted disabled persons but maintained an exception in cases of a "physical or mental disorder and behavior associated with the disorder that may pose, or has posed, a threat to the property, safety, or welfare of the alien or others." The exception is rooted in unfounded fears about psychiatric disability and violent behavior, and it is so vague and open to interpretation that arbitrary enforcement was inevitable.[17] The *Toronto Star*, for example, reported a case in 2013 of a Canadian denied entry because she had been briefly hospitalized a year and a half earlier for depression. Ellen Richardson, a paraplegic, was on her way to a Christmas cruise sailing from New York to the Caribbean, with a group organized by the March of Dimes, when she was told that "system checks" had flagged her as having had a "mental illness episode." She asked if she could telephone her own psychiatrist for approval, but instead the law required her to return to Toronto to get clearance from one of three Department of Homeland Security–approved doctors there, which of course put an end to her plans.[18] Richardson's experience was not unusual. The Psychiatric Patient Advocate Office in Toronto reported in 2011 that, in just that year, more than a dozen Canadians had reported being rejected at the border because of a record of mental illness, and that the office frequently fields calls from Canadians wishing to enter the United States but worried about being turned back on that account.[19]

The public-charge provision was also retained in the 1990 law, and disabled people continue to be excluded under it. Mark Weber argues that "effectively, the law allows consular officials to apply stereotypes in forecasting whether a person with a disabling condition is likely to be employable."[20] John Stanton has recounted the case of a Cuban extended family who applied to immigrate to the United States in 1995. They had family members who had been living in the country for many years and a sponsor who guaranteed them employment, thus satisfying the two primary admission criteria. Six of the group were granted visas, but two were refused because they were deaf and therefore likely to become public charges.[21] Kathrin S. Mautino, an attorney who specializes in immigration and nationality law, maintains that "those with physical disabilities meet real challenges under the immigration laws. Although most of the laws are facially neutral, many hidden restrictions remain." She cites instances, for example, in which the examining physician has found that a visa applicant's disabilities pose no threat, yet "nonetheless, the disability makes the officer nervous." Rather than challenging the medical finding directly, "consular and INS officers often look for other grounds" to justify exclusion, and "one of the most common is the public charge provision."[22] The Americans with Disabilities Act was necessary precisely because of common preconceptions that kept disabled citizens out of the workplace and public spaces. It is not surprising that immigration officials, who need not consider federal disability law, should act on similar assumptions.

As a prominent advocate of restriction wrote in 1930, "The necessity of the exclusion of the crippled, the blind, those who are likely to become public charges, and, of course, those with a criminal record is self evident."[23] The necessity has been treated as self-evident by scholars as well, such that discrimination against people with disabilities in the history of immigration law has gone largely unrecognized and unexamined. In 1973 the Children's Defense Fund, founded that year by Marian Wright Edelman, conducted a survey to identify the estimated 750,000 American children who were not attending school. Edelman expected to find large numbers of African Americans shut out from segregated school districts. What she found instead was that "handicapped kids were those seven hundred fifty thousand kids." She recalled that "we'd never thought of handicapped kids. But they're out there everywhere."[24] Historians could say the same. For a long time we never thought of disabled people, but they are everywhere, and our histories are defective without them.

Notes

Introduction

1. Before the federal government exercised authority over the border, some states (and, earlier, colonies) regulated their own borders. New York and Massachusetts in particular attempted to keep out paupers and other undesirables. See Roy L. Garis, *Immigration Restriction: A Study of the Opposition to and Regulation of Immigration in the United States* (New York: Macmillan, 1927), 16, 43–45, 75–77; Gerald L. Neuman, "The Lost Century of American Immigration Law (1776–1875)," *Columbia Law Review* 93, no. 8 (December 1, 1993): 1833–1901; Hidetaka Hirota, "The Moment of Transition: State Officials, the Federal Government, and the Formation of American Immigration Policy," *Journal of American History* 99, no. 4 (March 2013): 1092–1108.

2. United States Public Health Service (USPHS), *Regulations Governing the Medical Inspection of Aliens* (Washington, DC: Government Printing Office, 1917), 16–19; for records on the three freak-show performers, see National Archives, Records of the Immigration and Naturalization Service, Reports of Medical Inspectors in Philadelphia and New York, 1896–1903, RG 85, entry 2, pp. 241–44, dated December 19, 1899; "Three New Freaks in Town," *New York Times*, February 1, 1900. For more on Middlemiss, see chapter 3.

3. U.S. Bureau of Immigration, *Annual Report of the Commissioner of Immigration* (Washington, DC: Government Printing Office, 1907), 62.

4. Lucy E. Salyer, *Laws Harsh as Tigers: Chinese Immigrants and the Shaping of Modern Immigration Law* (Chapel Hill: University of North Carolina Press, 1995), 8, 15–16.

5. *Lennon v. Immigration and Naturalization Service*, 527 F.2d 187 (2d Cir. 1975).

6. See, for example, Eithne Luibhéie, *Entry Denied: Controlling Sexuality at the Border* (Minneapolis: University of Minnesota Press, 2002); Martha Gardner, *The Qualities of a Citizen: Women, Immigration, and Citizenship, 1870–1965* (Princeton, NJ: Princeton University Press, 2005); Jeanne D. Petit, *The Men and Women We Want: Gender, Race, and the Progressive Era Literacy Test Debate* (Rochester, NY: University of Rochester Press, 2010).

7. What attention has been given to disability and immigration has come mostly from nonhistorian disability scholars, such as Jay Dolmage, "Disabled upon Arrival: The Rhetorical Construction of Disability and Race at Ellis Island," *Cultural Critique* 77, no. 1 (2011): 24–69; Penny L. Richards, "Points of Entry: Disability and the Historical Geography of Immigration," *Disability Studies Quarterly* 24, no. 3 (June 15, 2004), http://dsq-sds.org/article/view/505;

James W. Trent, Jr., *Inventing the Feeble Mind: A History of Mental Retardation in the United States* (Berkeley and Los Angeles: University of California Press, 1994), 167–69.

8. Scholarly monographs and collections that have brought disability into some of these fields of history include: Henry Friedlander, *The Origins of Nazi Genocide: From Euthanasia to the Final Solution* (Chapel Hill: University of North Carolina Press, 1995); Martin S. Pernick, *The Black Stork: Eugenics and the Death of "Defective" Babies in American Medicine and Motion Pictures since 1915* (New York: Oxford University Press, 1996); David Gerber, ed., *Disabled Veterans in History* (Ann Arbor: University of Michigan Press, 2000); Paul K. Longmore and Lauri Umansky, eds., *The New Disability History: American Perspectives* (New York: New York University Press, 2001); David Serlin, *Replaceable You: Engineering the Body in Postwar America* (Chicago: University of Chicago Press, 2004); Jeffrey S. Reznick, *Healing the Nation: Soldiers and the Culture of Caregiving in Britain During the Great War* (New York: Manchester University Press, 2004); Beth Linker, *War's Waste: Rehabilitation in World War I America* (Chicago: University of Chicago Press, 2011); Nancy J. Hirschmann and Beth Linker, eds., Civil Disabilities: Citizenship, Membership, and Belonging (Philadelphia: University of Pennsylvania Press, 2015).

9. Douglas C. Baynton, "Disability and the Justification of Inequality in American History," in *The New Disability History: American Perspectives*, ed. Paul K. Longmore and Lauri Umansky (New York: New York University Press, 2001), 33–57.

10. Trent, Jr., *Inventing the Feeble Mind*, 141–44, 188–89; Allison C. Carey, *On the Margins of Citizenship: Intellectual Disability and Civil Rights in Twentieth-Century America* (Philadelphia: Temple University Press, 2009), 52–82; Philip Reilly, *The Surgical Solution: A History of Involuntary Sterilization in the United States* (Baltimore, MD: Johns Hopkins University Press, 1991); Pernick, *The* Black Stork.

11. Robert Buchanan, *Illusions of Equality: Deaf Americans in School and Factory, 1850–1950* (Washington, DC: Gallaudet University Press, 1999), 20–36; Douglas C. Baynton, *Forbidden Signs: American Culture and the Campaign against Sign Language* (Chicago: University of Chicago Press, 1996). On "ashamed to sign in public," see Douglas C. Baynton, Jack R. Gannon, and Jean Lindquist Bergey, *Through Deaf Eyes: A Photographic History of an American Community* (Washington, DC: Gallaudet University Press, 2007), 117.

12. Susan M. Schweik, *The Ugly Laws: Disability in Public* (New York: New York University Press, 2009).

13. Paul K. Longmore, *Why I Burned My Book and Other Essays on Disability* (Philadelphia: Temple University Press, 2003), 36–37.

14. Michael Omi and Howard Winant, "Racial Formation in the United States," in *The Idea of Race*, ed. Robert Bernasconi and Tommy Lee Lott (Cambridge, MA: Hackett, 2000), 183.

15. Perhaps the most cogent and illuminating discussion of European "races" and the twentieth-century shift toward a language of "ethnicity" can be found in David R. Roediger's *Working toward Whiteness: How America's Immigrants Became White; The Strange Journey from Ellis Island to the Suburbs* (New York: Basic Books, 2005), 14–37. Roediger notes that the concept of "ethnicity" did not begin to replace "race" when discussing Europeans until the 1930s and argues that for historians, "transporting ethnicity backward in time . . . puts a layer of mystification between us and the past" (33). See also Matthew Frye Jacobson, *Whiteness of a Different Color: European Immigrants and the Alchemy of Race* (Cambridge, MA: Harvard University Press, 1998), 68–69.

16. Mae M. Ngai, *Impossible Subjects: Illegal Aliens and the Making of Modern America* (Princeton, NJ: Princeton University Press, 2004), 26–27.

17. John Higham, *Strangers in the Land Patterns of American Nativism, 1860–1925*, 2nd ed. (New Brunswick, NJ: Rutgers University Press, 2002).

Chapter One

1. Simo Vehmas, "Live and Let Die? Disability in Bioethics," *New Review of Bioethics* 1, no. 1 (November 2003): 145–57; Ayo Wahlberg, "Reproductive Medicine and the Concept of 'Quality,'" *Clinical Ethics* 3, no. 4 (December 2008): 160–63; Jackie Leach Scully, "Disability and Genetics in the Era of Genomic Medicine," *Nature* 9 (October 2008): 797–802. Portions of chapter 1 are reprinted by permission of the publisher from "Defect: A Selective Reinterpretation of American Immigration History," in Nancy Hirschmann and Beth Linker, eds., *Civil Disabilities: Citizenship, Membership, and Belonging*, University of Pennsylvania Press, 2015, pp. 45–64. Copyright © University of Pennsylvania Press.

2. Charles Darwin, *Variation of Animals and Plants under Domestication*, vol. 1 (New York: D. Appleton, 1915), 6.

3. Charles Darwin, *The Descent of Man, and Selection in Relation to Sex* (New York: D. Appleton, 1871), 161–62.

4. Euthanasia was an extreme position in the American eugenics movement (though not in the German movement by the 1930s) and never sanctioned in public policy or law. Martin S. Pernick found that although most eugenicists had long taken a public position against euthanasia, few openly spoke against the physician Harry Haiselden when he advocated and practiced euthanasia of disabled infants; see *The Black Stork: Eugenics and the Death of "Defective" Babies in American Medicine and Motion Pictures since 1915* (New York: Oxford University Press, 1996). James W. Trent Jr. writes that "Charles Eliot Norton, professor and former president of Harvard University and editor of the *North American Review*, advocated for 'the painless destruction' of insane and deficient minds"; *Inventing the Feeble Mind: A History of Mental Retardation in the United States* (Berkeley and Los Angeles: University of California Press, 1994), 134. Paul A. Lombardo found that Oliver Wendell Holmes had, in private letters, advocated euthanasia for disabled infants; *Three Generations, No Imbeciles: Eugenics, the Supreme Court, and Buck v. Bell* (Baltimore, MD: Johns Hopkins University Press, 2008), 165.

5. Robert DeCourcy Ward, *Crisis in Our Immigration Policy*, Publications of the Immigration Restriction League 61 (Boston: The League, 1913): 7. Henry Fairfield Osborn, letter to the editor, *New York Evening Journal*, June 19, 1911.

6. James W. Trent Jr., *Inventing the Feeble Mind: A History of Mental Retardation in the United States* (Berkeley and Los Angeles: University of California Press, 1994), 131–83; Philip R. Reilly, *The Surgical Solution: A History of Involuntary Sterilization in the United States* (Baltimore, MD: Johns Hopkins University Press, 1991).

7. Harry H. Laughlin, "Eugenical Sterilization in the United States," *Social Hygiene* 6, no. 4 (October 1920): 530. Eugenicists sometimes spoke of four tracks: education, restriction, segregation, and sterilization; see Michael Rembis, *Defining Deviance: Sex, Science, and Delinquent Girls, 1890–1960* (Urbana: University of Illinois Press, 2013), 3.

8. Albert Edward Wiggam, *The New Decalogue of Science* (Garden City, NY: Garden City, 1925), 110–11.

9. American Eugenics Society, *A Eugenics Catechism* (New Haven, CT: The Society, 1926), 10.

10. Charles Richmond Henderson, "Are Modern Industry and City Life Unfavorable to the Family?," *American Economic Association Quarterly*, 3rd ser., 10, no. 1 (1909): 228.

11. Harry Hamilton Laughlin, *Eugenical Sterilization in the United States* (Chicago: Psychopathic Laboratory of the Municipal Court of Chicago, 1922), 339.

12. Pernick, *The Black Stork*, 95–96.

13. Susan Burch and Hannah Joyner, *Unspeakable: The Story of Junius Wilson* (Chapel Hill: University of North Carolina Press, 2007).

14. Daniel J. Kevles, *In the Name of Eugenics: Genetics and the Uses of Human Heredity* (Berkeley and Los Angeles: University of California Press, 1985), 99–100.

15. "Putting Our Immigrants through the Sieve: Government Stands as 'Doctor of Eugenics' at Portals of Nation," Portland Oregonian, September 28, 1913. Victor Elliott, "Eugenicists Declare Immigrants Peril," *Fort Worth Star-Telegram*, July 5, 1914.

16. Eugene Solomon Talbot, *Degeneracy: Its Causes, Signs, and Results* (London: W. Scott, 1898), 19.

17. George Lydston, *The Diseases of Society: The Vice and Crime Problem* (Philadelphia: J. B. Lippincott, 1908), 93.

18. Francis Galton, "Studies in Eugenics," *American Journal of Sociology* 11, no. 1 (July 1905): 23.

19. "Significance of a Sound Physique," *New York Times*, April 18, 1909.

20. Walter E. Fernald, "The Imbecile with Criminal Instincts," *American Journal of Psychiatry* 65, no. 4 (April 1, 1909): 747. See also John B. Weber and Charles Stewart Smith, "Our National Dumping-Ground. A Study of Immigration," *North American Review* 154, no. 4 (April 1892): 425, which referred to "paupers, criminals, or other defectives."

21. Letter from Victor Safford to the Commissioner, May 16, 1906, Immigration Restriction League Records, US 10583.9.8–US 10587.43, box 4, folder Safford, M. Victor, "Definitions of Various Medical Terms Used in Medical Certificates, 1906," Houghton Library, Harvard University.

22. Martin W. Barr, *Mental Defectives: Their History, Treatment, and Training* (Philadelphia: P. Blakiston's, 1904), 100–101. See also Nicole Hahn Rafter, Creating Born Criminals (Urbana: University of Illinois Press, 1997), 37.

23. "The Study of the Criminal," *Boston Medical and Surgical Journal* 125 (July–December 1891): 579.

24. Laws restricting Asian immigration have usually been treated separately, as they constituted a separate body of legislation. See Mae M. Ngai, *Impossible Subjects: Illegal Aliens and the Making of Modern America* (Princeton, NJ: Princeton University Press, 2004), 18; Lucy Salyer, *Laws Harsh as Tigers: Chinese Immigrants and the Shaping of Modern Immigration Law* (Chapel Hill: University of North Carolina Press, 1995), viii, 32; Roger Daniels, *Guarding the Golden Door: American Immigration Policy and Immigrants since 1882* (New York: Hill & Wang, 2004), 3, 19–26.

25. Philip Taylor, *The Distant Magnet: European Immigration to the U.S.A.* (New York: Harper & Row, 1971), 241–42.

26. Roger Daniels, *Coming to America: A History of Immigration and Ethnicity in American Life* (New York: HarperCollins, 1990), 274, 278–79.

27. Bill Ong Hing, *Defining America through Immigration Policy* (Philadelphia: Temple University Press, 2012), 7. Howard Markel and Alexandra Minna Stern have noted that during the years 1891 to 1924 the percentage of exclusions that cited medical reasons increased dramatically, and that this was driven by exclusions for disability rather than infectious disease; see Markel and Stern, "The Foreignness of Germs: The Persistent Association of Immigrants and Disease in American Society," *Milbank Quarterly* 80, no. 4 (January 1, 2002): 763–64.

28. Claudia Goldin, "The Political Economy of Immigration Restriction in the United States, 1890 to 1921," in *The Regulated Economy: A Historical Approach to Political Economy*, ed.

Claudia Dale Goldin and Gary D. Libecap (Chicago: University of Chicago Press, 1994), 223. See also Ian Robert Dowbiggin, *Keeping America Sane: Psychiatry and Eugenics in the United States and Canada, 1880–1940* (Ithaca, NY: Cornell University Press, 1997), 192.

29. Hasia R. Diner, *The Jews of the United States, 1654 to 2000* (Berkeley and Los Angeles: University of California Press, 2004), 75.

30. Randall Hansen and Desmond King, "Eugenic Ideas, Political Interests, and Policy Variance: Immigration and Sterilization Policy in Britain and the U.S.," *World Politics* 53, no. 2 (January 2001): 242.

31. Ngai, *Impossible Subjects*, 17. Peter Schrag qualified the claim slightly, writing that the 1921 act "put a permanent end to the nation's history of virtually open borders and set a radically new pattern for the immigration ideology"; *Not Fit for Our Society: Immigration and Nativism in America* (Berkeley and Los Angeles: University of California Press, 2010), 116.

32. Michael Lemay and Elliott Robert Barkan, eds., *U.S. Immigration and Naturalization Laws and Issues: A Documentary History* (Westport, CT: Greenwood Press, 1999), 41.

33. Keith Fitzgerald, *The Face of the Nation: Immigration, the State, and the National Identity* (Palo Alto, CA: Stanford University Press, 1996), 19, 116–17, 129; emphasis added.

34. *United States Statutes at Large*, vol. 22 (Washington, DC: Government Printing Office, 1883), 214 (hereafter abbreviated as *U.S. Statutes*); *U.S. Statutes*, vol. 26 (1891), 1084; *U.S. Statutes*, vol. 34 (1907), 898–99; *U.S. Statutes*, vol. 32 (1903), 1213. Earlier legislation in 1875 excluded criminals and prostitutes.

35. *Biological Aspects of Immigration: Hearings before the Committee on Immigration and Naturalization*, H.R., 66th Cong., 2d Sess., at 20 (April 16–17, 1920) (statement of Harry H. Laughlin). See also Desmond King, *Making Americans: Immigration, Race, and the Origins of the Diverse Democracy* (Cambridge, MA: Harvard University Press, 2009), 176.

36. United States Public Health Service, *Regulations Governing the Medical Inspection of Aliens* (Washington, DC: U.S. Government Printing Office, 1917), 30–31.

37. U.S. Bureau of Immigration, *Annual Report of the Commissioner-General of Immigration* (Washington, DC: Government Printing Office, 1907), 62 (hereafter cited by title and year alone).

38. William Williams, in *Annual Report of the Commissioner-General of Immigration* (1910), 135.

39. *Annual Report of the Commissioner-General of Immigration* (1912), 123.

40. *Annual Report of the Commissioner-General of Immigration* (1919), 15–16.

41. John Higham, *Strangers in the Land: Patterns of American Nativism, 1860–1925* (New Brunswick, NJ: Rutgers University Press, 1998), 99, 186–87. Higham did not always use "restriction" consistently, occasionally referring to laws before the 1917 act as "restriction," but his emphasis remains clear. In addition, after describing the 1917 law as restrictive, he later referred to Wilson's pocket veto of the 1921 quota bill as "the inarticulate swan song of unrestricted immigration" (311), perhaps because, as he noted, the literacy test was less effective than hoped owing to recent improvements in European literacy rates (308).

42. *U.S. Statutes*, vol. 34 (1907), 898–99.

43. Higham, *Strangers in the Land*, 101, 162–63, 186–87, 202–4. Higham also suggested that eugenicists "in their scientific capacity" came to "logical and consistent" conclusions when focusing on individual immigrants, but "when they generalized the defects of individual immigrants into those of whole ethnic groups, their science deserted them" (152–53).

44. Marion T. Bennett, *American Immigration Policies: A History* (Washington, DC: Public Affairs Press, 1963), 15, 29. Despite his periodization, Bennett also used "restriction" loosely

at times to refer to all immigration laws since 1875. The immigration law scholar Bina Kalola similarly wrote that the laws "began as a selection process and developed [in the 1920s] into a means of restricting the mass flow of immigrants"; "Immigration Laws and the Immigrant Woman: 1885–1924," *Georgetown Immigration Law Journal* 11, no. 553 (Spring 1997): 553. Eithne Luibhéie also writes that "selective immigration" was "official United States policy"; Luibhéie, *Entry Denied: Controlling Sexuality at the Border* (Minneapolis: University of Minnesota Press, 2002), 2, 9.

45. Edward Prince Hutchinson, *Legislative History of American Immigration Policy, 1798–1965* (Philadelphia: University of Pennsylvania Press, 1981), 167.

46. Roger Daniels, *Coming to America: A History of Immigration and Ethnicity in American Life* (New York: HarperCollins, 1990), 274. In a later work, Daniels characterized Hutchinson's depiction of a shift "from regulation to restriction" as a "false distinction." His argument, however, was based on the Chinese Exclusion Act of 1882 rather than the laws concerning defectives. Daniels, *Guarding the Golden Door*, 3, 17–26.

47. Kenneth M. Ludmerer, *Genetics and American Society: A Historical Appraisal* (Baltimore, MD: Johns Hopkins University Press, 1972), 96–97. Similarly, Daniel J. Kevles argued that "economic factors . . . tended to dominate the debate through 1921"; Kevles, *In the Name of Eugenics: Genetics and the Uses of Human Heredity* (Berkeley and Los Angeles: University of California Press, 1985), 96.

48. See Paul K. Longmore and David Goldberger, "The League of the Physically Handicapped and the Great Depression: A Case Study in the New Disability History," *Journal of American History* 87, no. 3 (December 2000): 888–922.

49. Deirdre M. Moloney, "Women, Sexual Morality, and Economic Dependency in Early U.S. Deportation Policy," *Journal of Women's History* 18, no. 2 (Summer 2006): 98.

50. National Archives, Records of the Immigration and Naturalization Service (hereafter cited as Records of the INS), RG 85, entry 7, file 48,599/4.

51. National Archives, Records of the INS, RG 85, entry 7, file 49833–2.

52. National Archives, Records of the INS, RG 85, Accession 60A600, file 53,700/974.

53. National Archives, Records of the INS, RG 85, Accession 60A600, file 51,806–16.

54. Martha C. Nussbaum, *Hiding from Humanity: Disgust, Shame, and the Law* (Princeton, NJ: Princeton University Press, 2004), 307–8.

55. Quoted in Brad Byrom, "A Pupil and a Patient: Hospital Schools in Progressive America," in Longmore and Umansky, *New Disability History*, 138–39.

56. Robert M. Buchanan, *Illusions of Equality: Deaf Americans in School and Factory, 1850–1950* (Washington, DC: Gallaudet University Press, 1999), 75–76; Sarah Frances Rose, "No Right to be Idle: The Invention of Disability, 1850–1930" (Ph.D. diss., University of Chicago, 2008), 142–57.

57. Gertrude R. Stein, "Placement Technique in the Employment Work of the Red Cross for Crippled and Disabled Men," in *Publications of the Red Cross Institute for Crippled and Disabled Men*, series 1, ed. Douglas C. McMurtrie (New York: Red Cross Institute, 1918), 3–4.

58. Frank B. Gilbreth and Lillian Moller Gilbreth, *Motion Study for the Handicapped* (London: Routledge, 1920), 27, 135–36. Salyer, in her fine history of the laws regarding Chinese immigration, notes briefly that a concern about public charges "motivated these policies in part," but that "increasingly in the early 1900s, lawmakers and nativists justified the policy on biological grounds, accepting the eugenicists' argument that by excluding mental and physical defectives, they would prevent the deterioration of the American mind and body." Salyer, *Laws Harsh as Tigers*, 130.

59. Evidence for this assertion can be found in chapter 3, where I explore the issue of dependency at greater length.

60. See, for example, Richmond Mayo-Smith, *Emigration and Immigration: A Study in Social Science* (New York: Charles Scribner's Sons, 1890), 276; *Annual Report of the Commissioner-General of Immigration* (1897), 4–5; Frederick A. Bushee, "Ethnic Factors in the Population of Boston," *Publications of the American Economic Association*, 3rd ser., 4, no. 2 (May 1903): 12.

61. United States, *Annual Report of the Superintendent of Immigration* (Washington, DC: Government Printing Office, 1893), 8, 11 (hereafter cited by title and year alone).

62. *Annual Report of the Superintendent of Immigration* (1893), 10–11.

63. *Annual Report of the Superintendent of Immigration* (1894), 12–13. The prediction was vindicated the following year, the decline being steepest from "countries usually furnishing many of the most undesirable immigrants;" *Annual Report of the Commissioner-General of Immigration* (1895), 13.

64. *Annual Report of the Commissioner-General of Immigration* (1897), 4–5.

65. Weber and Smith, "Our National Dumping-Ground," 437–38. William E. Chandler, "Shall Immigration Be Suspended?," *North American Review* 156, no. 434 (January 1893): 4, 6–8. Joseph H. Senner, "How We Restrict Immigration," *North American Review* 158, no. 449 (April 1894): 498.

66. Henry Sidney Everett, "Immigration and Naturalization," *Atlantic Monthly*, March 1895, 352; *Congressional Record—Senate*, 54th Cong., 1st Sess. (March 16, 1896), vol. 28, part 3:2817; emphasis added.

67. "President Grover Cleveland's Veto Message of the Immigrant Literacy Test Bill (March 2, 1897)," in Michael C. LeMay and Elliott Robert Barkan, *U.S. Immigration and Naturalization Laws and Issues: A Documentary History* (Westport, CT: Greenwood Press, 1999), 80–82.

68. Committee on Immigration and Naturalization, Immigration of Aliens into the United States: Views of the Minority, H.R. Rep. No. 95, pt. 2, 64th Cong., at 1 (February 8, 1916).

69. "President Woodrow Wilson's Veto Message of the Immigrant Literacy Test Bill (January 28, 1915)," in Michael C. LeMay and Elliott Robert Barkan, *U.S. Immigration and Naturalization Laws and Issues: A Documentary History* (Westport, CT: Greenwood Press, 1999), 107–8. On the other hand, Harry Laughlin, in testimony before Congress in 1924, maintained that "thus far our policy has been largely negatively restrictive rather than positively selective"; *Europe as an Immigrant-Exporting Continent and the United States as an Immigrant-Receiving Nation: Hearings before the Committee on Immigration and Naturalization*, 68th Cong., H.R. Serial 5-A, at 1238 (March 8, 1924) (statement of Dr. Harry H. Laughlin).

70. Immigration Restriction League, "Constitution of the Immigration Restriction League" (Boston: The League, [1894 or 1895]): 1.

71. Immigration Restriction League, "Resolutions Relative to the Restriction of Immigration," in "Scrapbook: Immigration Restriction League, 1896–1898," vol. 1, Series III, Records (MS Am 2245), Houghton Library, Harvard University.

72. Immigration Restriction League, *The Reading Test: Why It Should Be Adopted*, Publications of the Immigration Restriction League 63 (Boston: The League, 1914), 5.

73. Prescott Hall "Selection of Immigration," *Annals of the American Academy of Political and Social Science* 24, no. 1 (July 1904): 175, 182.

74. Robert DeCourcy Ward, "The Immigration Problem," *Charities* 12 (February 1904): 138, 150. See also Robert DeCourcy Ward, "National Eugenics in Relation to Immigration," *North American Review* 192 (1910): 64; Winfield S. Alcott, "The Modern Problem of Immigration," *New England Magazine*, July 1906, 553, 560, 562 (the literacy test was "the only immediate

practicable means of reducing the immense volume of immigration as well as improving the average quality").

75. *Annual Report of the Commissioner-General of Immigration* (1903), 61.

76. Higham, *Strangers in the Land*, 101.

77. Barbara Miller Solomon, *Ancestors and Immigrants: A Changing New England Tradition* (Boston: Northeastern University Press, 1989), 102. Other historians have followed and cited Solomon's argument in their accounts of the literacy test; see, for example, Daniels, *Guarding the Golden Door*, 31; King, *Making Americans*, 53.

78. See, for example, Hall, "Selection of Immigration," *Annals*, 179, 182.

79. Immigration Restriction League, *Brief in Favor of the Illiteracy Test*, Publications of the Immigration Restriction League 56 (Boston: The League, 1910), 4.

80. Prescott Hall, "Immigration and the Educational Test," *North American Review* 165, no. 491 (October 1897): 395.

81. Robert DeCourcy Ward, *Crisis in Our Immigration Policy*, Publications of the Immigration Restriction League 61 (Boston: The League, 1913), 6 (also in Ward, "Race Betterment and Our Immigration Laws," *Official Proceedings of the National Conference on Race Betterment* [Battle Creek, MI: Race Betterment Foundation, 1914], 543); Ward, "The Immigration Problem," 140. Similarly, President Theodore Roosevelt deprecated drawing race lines in regulating immigration in his message of 1905 when he said, "We cannot afford to pay heed to whether he is of one creed or another, of one nation or another. . . . What we should desire to find out is the individual quality of the individual man"; quoted in Max J. Kohler, "Restriction of Immigration—Discussion," *American Economic Review* 2, no. 1 (March 1912): 77.

82. Ward, "Immigration Problem," 147–49.

83. Robert DeCourcy Ward, "Our Immigration Laws from the Viewpoint of National Eugenics," *National Geographic Magazine*, January 1912, 39.

84. Leila Zenderland, "The Parable of the Kallikak Family," in *Mental Retardation in America: A Historical Reader*, ed. Steven Noll and James W. Trent (New York: New York University Press, 2004), 177–78.

85. *Annual Report of the Commissioner-General of Immigration* (1910), 5.

86. Montaville Flowers, *The Japanese Conquest of American Opinion* (New York: George H. Doran, 1917), 261.

87. Charles B. Davenport, "Influence of Heredity on Human Society," *Annals of the American Academy of Political and Social Science* 34 (1909): 21.

88. Charles Benedict Davenport, *Heredity in Relation to Eugenics* (New York: Henry Holt, 1911), 222–24.

89. "Unsocial Blood-Lines Often Brought In with Immigrant Germ Plasm, Mr. Davenport Says," *New York Times*, June 9, 1913.

90. "Minutes of Meeting of Committee on Immigration of the Eugenics Research Association Held at the Harvard Club, New York City, at 11 A.M., February 25, 1920," Charles B. Davenport Papers, ca. 1903–1940, B: D27; Series IIb. Cold Spring Harbor Series; Folder: Eugenics Research Association—Committee on Immigration, American Philosophical Society Library, Philadelphia.

91. Quoted in Madison Grant, *The Passing of the Great Race: Or, The Racial Basis of European History* (New York: Charles Scribner's Sons, 1921), 303 n. 110. Even Grant, who was indeed obsessed with the racial origins of immigrants, described the objectionable ones as "the weak, the broken, and the mentally crippled of all races drawn from the lowest stratum of the Mediter-

ranean and the Balkins [*sic*]"; the superiority of northern Europeans he ascribed to the "rigid elimination of defectives" due to the harsh climate; see pp. 89, 170.

92. "Like-Minded or Well-Born?" *New York Times*, February 10, 1924.

93. "Eugenics and Immigration: Johnson Bill Favored Only as a Step Forward," *New York Times*, February 16, 1924.

94. *Annual Report of the Commissioner-General of Immigration* (1898), 32. Poor physique is discussed in Deirdre M. Moloney, *National Insecurities: Immigrants and U.S. Deportation Policy since 1882* (Chapel Hill: University of North Carolina Press, 2012), 283 n. 47; and Amy L. Fairchild, *Science at the Borders: Immigrant Medical Inspection and the Shaping of the Modern Industrial Labor Force* (Baltimore, MD: Johns Hopkins University Press, 2003), 165–70. Moloney argues that poor physique was intended primarily to exclude Jews and Eastern Europeans, but the evidence shows that to be only part of the motivation.

95. "Immigration Problems before the Nation," *New York Times*, May 24, 1903. William Williams, "Ellis Island Station," in *Annual Report of the Commissioner-General of Immigration* (1904), 105–6; William Williams, "The New Immigration: Some Unfavorable Features and Possible Remedies," *Proceedings of the National Conference of Charities and Correction* 33 (May 1906): 7–8; William Williams, "Remarks on Immigration," unpublished address to the senior class of Princeton University (1904), in William Williams Papers, New York Public Library; quoted in Alan M. Kraut, *Silent Travelers: Germs, Genes, and the "Immigrant Menace"* (Baltimore, MD: Johns Hopkins University Press, 1995), 68, 298 n. 57.

96. Allan J. McLaughlin, "Immigration and the Public Health," *Public Health Papers and Reports* (Columbus, OH: Press of Fred J. Heer, 1904): 224–31. Allan J. McLaughlin, "The American's Distrust of the Immigrant," *Popular Science Monthly*, January 1903, 230–36.

97. McLaughlin article reprinted in *Popular Science Monthly*, January 1904, 232–38; *Public Opinion* 1 (September 23, 1905): 3–4; Robert M. La Follette, *The Making of America*, vol. 10 (Chicago: De Bower, Chapline, 1907), 119–27. See also Robert DeCourcy Ward, "National Eugenics in Relation to Immigration," *North American Review* 192 (July 1910): 67. "Immigration Problems before the Nation," *New York Times*, May 24, 1903.

98. Allan McLaughlin, "Social and Political Effects of Immigration," *Popular Science Monthly*, January 1905, 243–55; "How Immigrants Are Inspected," *Popular Science Monthly*, February 1905, 357–61; "The Problem of Immigration," *Popular Science Monthly*, April 1905, 531–37.

99. Allan McLaughlin, "The Problem of Immigration," *Popular Science Monthly*, April 1905, 532; emphasis added. The Surgeon General in 1907 justified this view at a meeting of immigration officials: "If one man comes in healthy, and another has a deformity or a horrible disease on his face, the latter would not stand a chance of getting work. They would accept the good man. Does not that affect his ability to get employment?" "Conference to Consider Medical Examination of Immigrants [minutes]," February 8, 1907, National Archives, Records of the INS, RG 85, entry 9, file 51490/19, p. 7.

100. Letter from Robert DeCourcy Ward to Frank P. Sargent, Commissioner-General of Immigration, Washington, DC, January 11, 1905, National Archives, Records of the INS, RG 85, entry 9, File 51490/19.

101. Letter, March 28, 1905, National Archives, Records of the INS, RG 85, entry 9, file 51490/19. See also letter from F. P. Sargent, Commissioner-General of the Bureau of Immigration, to the Commissioner of Immigration on Ellis Island, April 17, 1905, National Archives, Records of the Public Health Service, RG 90, entry 10, file 219.

102. Letter, March 28, 1905, National Archives, Records of the INS, RG 85, entry 9, file 51490/19.

103. "Immigration Rule Hits Jews: Reasons Why They Appear of 'Poor Physique,'" *Boston Evening Transcript*, April 12, 1905, 14.

104. Letter, April 12, 1905, National Archives, Records of the INS, RG 85, entry 9, file 51490/19. Ward wrote again a week later to again insist on the necessity to "*specifically* exclude persons of poor physique," as they would become "fathers and mothers of future American children." Letter, April 19, 1905, National Archives, Records of the INS, RG 85, entry 9, file 51490/19.

105. Memorandum from F. P. Sargent, Commissioner-General of the Bureau of Immigration, April 17, 1905, National Archives, Records of the Public Health Service, RG 90, entry 10, file 219. Letter from Sargent to Ward, April 25, 1905, National Archives, Records of the INS, RG 85, entry 9, file 51490/19.

106. Letter from Acting Commissioner General Frank Larned to Walter Wyman, Surgeon General, December 22, 1906, National Archives, Records of the INS, RG 85, entry 9, file 48,462/3.

107. National Archives, Records of the INS, RG 85, entry 7, file 49,968/4. Abram Vossen Goodman, "Adolphus S. Solomons and Clara Barton: A Forgotten Chapter in the Early Years of the American Red Cross," *American Jewish Historical Quarterly* 59, no. 3 (March 1970): 331–56; *Jewish Encyclopedia*, vol. 11 (New York: Funk & Wagnall's, 1925), 459; Dumas Malone, ed., *Dictionary of American Biography*, vol. 17 (New York: Charles Scribner's Sons, 1943), 393–94.

108. In 1924 Ward wrote that "appeals are constantly made on behalf of every class of alien specified in the law as excludable. Immigrant 'aid' and charitable societies; immigrant lawyers who prey on the alien; relatives and friends of those liable to be debarred; senators and congressmen of the United States and city and state politicians more interested in securing a few votes from their foreign-born constituents than in the future character of our race; misguided sentimentalists, appeal in behalf of the detained alien." Robert DeCourcy Ward, "Higher Mental and Physical Standards for Immigrants," *Scientific Monthly* 19 (November 1924): 536. The Ellis Island commissioner, Robert Watchorn, complained that appeals were too frequently granted, overturning the conclusions of local inspectors; "Is There an Immigrant Peril?," *National Civic Federation Review* 2, no. 3 (June 1905): 7.

109. National Archives, Records of the INS, RG 85, entry 9, file 51497/4.

110. Letter, March 14, 1906, National Archives, Records of the INS, RG 85, entry 9, file 48,462/3. See also Fairchild, *Science at the Borders*, 165–70. Howard Markel notes that Prescott Hall drew a connection between poor physique and susceptibility to disease; see *When Germs Travel: Six Major Epidemics That Have Invaded America since 1900 and the Fears They Have Unleashed* (New York: Pantheon Books, 2004), 35.

111. Letter, March 30, 1906, National Archives, Records of the INS, RG 85, entry 9, file 48,462/3.

112. Letter, December 22, 1906, National Archives, Records of the INS, RG 85, entry 9, file 48,462/3. The letter was signed by the Acting Commissioner General, Frank Larned, during Sargent's temporary absence, but Larned was acting on Sargent's behalf.

113. "Checks upon Immigration," *New York Times*, January 17, 1907; *U.S. Statutes*, vol. 26, 1084; *U.S. Statutes*, vol. 34, 899.

114. Letter from Prescott F. Hall to Walter Wyman, March 30, 1908, National Archives, Records of the INS, RG 90, entry 10, file 219. Unfortunately for both Hall and later historians, this suggestion was apparently never acted upon.

115. *Annual Report of the Commissioner-General of Immigration* (1910), 5.

116. Memorandum, February 10, 1914, National Archives, Records of the INS, Subject Correspondence 1906–1932, RG 85, entry 9, file 53139/10-M.

117. Nathan Bijur, "Is There an Immigrant Peril?," *National Civic Federation Review* 2, no. 3 (June 1905): 2. Bijur took his figures from Roland P. Falkner, "Some Aspects of the Immigration Problem," *Political Science Quarterly* 19, no. 1 (March 1904): 32–49.

118. W. F. Willcox, "Restriction of Immigration—Discussion," *American Economic Review* 2, no. 1 (March 1912): 68.

119. Excerpted from the *Rochester Chronicle*, in the *National Civic Federation Review* 2, no. 3 (June 1905): 20.

120. See, for example, William Williams, *Annual Report of 1908*, excerpted in *Annual Report of the Commissioner-General of Immigration* (1908), 138. James J. Davis, *Selective Immigration* (St. Paul, MN: Scott Mitchell, 1925), 47. *Annual Report of the Commissioner-General of Immigration* (1915), 10–11. Robert DeCourcy Ward, "Higher Mental and Physical Standards for Immigrants," *Scientific Monthly* 19 (November 1924): 533, 535.

121. Letter, June 10, 1911, Immigration Restriction League Records, US10583.9.8—US10587.43, box 2 Correspondence: Fe—Q, folder Fisk, Arthur L., Houghton Library, Harvard University.

122. Irving Fisher, "Impending Problems of Eugenics," *Scientific Monthly* 13, no. 3 (September 1921): 214, 227–30.

123. Robert DeCourcy Ward, *Crisis in Our Immigration Policy*, Publications of the Immigration Restriction League 61 (Boston: The League, 1913), 8.

124. *Physical Examination of Immigrants: Hearings before the Committee on Immigration and Naturalization*, H.R. 66th Cong., 3d Sess., at 11 (January 11, 1921).

125. Alfred C. Reed, "The Medical Side of Immigration," *Popular Science Monthly*, April 1912, 389–90; "Going through Ellis Island," *Popular Science Monthly*, January 1913, 8–9; "Immigration and the Public Health," *Popular Science Monthly*, October 1913, 325.

126. Thomas Wray Grayson, "The Effect of the Modern Immigrant on Our Industrial Centers," in American Academy of Medicine, *Medical Problems of Immigration: Being the Papers and Their Discussion Presented at the XXXVII Annual Meeting of the American Academy of Medicine, Held at Atlantic City, June 1, 1912* (Easton, PA: American Academy of Medicine, 1913), 103, 107–9.

127. Frederick A. Bushee, "Ethnic Factors in the Population of Boston," *Publications of the American Economic Association*, 3rd ser., 4, no. 2 (May 1903): 82–83, 150.

128. Thomas W. Salmon, "Immigration and the Mixture of Races in Relation to the Mental Health of the Nation," in *The Modern Treatment of Nervous and Mental Diseases*, ed. William Alanson White and Smith Ely Jelliffe (Philadelphia: Lee & Febiger, 1913), 275.

129. Edward Alsworth Ross, *The Old World in the New: The Significance of Past and Present Immigration to the American People* (New York: Century, 1914), 285–90.

130. Joseph G. Wilson, "A Study in Jewish Psychopathology," *Popular Science Monthly*, March 1913, 264–66, 271.

131. *Biological Aspects of Immigration: Hearings before the United States House Committee on Immigration and Naturalization*, 66th Cong., 2d Sess., at 3–4 (April 16–17, 1920) (statement of Harry H. Laughlin).

132. *Biological Aspects of Immigration*, 5, 13–15, 17.

133. *Analysis of America's Melting Pot: Hearings before the Committee on Immigration and Naturalization*, 67th Cong., 3d Sess., H.R. Serial 7-C, at 733, 752 (November 21, 1922) (statement of Harry H. Laughlin). Desmond King sees race as Laughlin's paramount concern but also notes that Laughlin believed the United States capable of assimilating northwestern Europeans "only if carefully selected as to inborn family qualities"; *Making Americans*, 134–35.

134. Madison Grant et al., "Third Report of the Sub-Committee on Selective Immigration of the Eugenics Committee of the United States of America The Examination of Immigrants Overseas, as an Additional Safeguard in the Processes of Enforcing American Immigration Policy," *Journal of Heredity* 16, no. 8 (August 1, 1925): 294–95. See also Nancy Ordover, *American Eugenics: Race, Queer Anatomy, and the Science of Nationalism* (Minneapolis: University of Minnesota Press, 2003), 31.

135. Robert DeCourcy Ward, "Higher Mental and Physical Standards for Immigrants," *Scientific Monthly* 19 (November 1924): 533, 535. The Harvard biologist Edward M. East bluntly stated what was still in 1930 the essential equation for restrictionists: "If in the future the proportion of people of Grades A and B increases, the nation will prosper; while if the proportion of people of Grades D and E increases, the nation will decay." He concluded that the "eugenic ideals" embodied in selective reproduction and selective immigration were "the sole and final means of keeping a nation from deterioration." Edward M. East, "Population Pressure and Immigration," in *The Alien in Our Midst; or, Selling Our Birthright for a Mess of Industrial Pottage*, ed. Madison Grant and Charles Steward Davison (New York: Galton, 1930), 93, 97.

136. Glenn Frank, "A Sensible Immigration Policy," *Century Magazine*, May 1924, 137–39.

137. Quoted in "1890 Census Urged as Immigrant Base: Eugenics Committee Report," *New York Times*, January 7, 1924.

138. *Europe as an Immigrant-Exporting Continent and the United States as an Immigrant-Receiving Nation, Hearings before the Committee on Immigration and Naturalization*, 68th Congress, 1st Sess., H.R. Serial 5-A, at 1238, 1278 (March 8, 1924) (statement of Dr. Harry H. Laughlin).

139. Davis, *Selective Immigration*, 33, 86. National quotas were removed in 1965. Most of the restrictions that applied specifically to disability were removed from the law in 1990. Disabled immigrants can still be denied an entry visa on the basis of the "likely to become a public charge" clause today, but because records are not kept of visa denials, it is not known how often this occurs. See John F. Stanton, "The Immigration Laws from a Disability Perspective: Where We Were, Where We Are, Where We Should Be," *Georgetown Immigration Law Journal* 10 (Spring 1996): 451.

140. Salyer, *Laws Harsh as Tigers*, 15–16. Matthew Frye Jacobson notes the connections between race and defect but argues that the restriction of disabled immigrants "rested upon racial distinctions." As I argue in this chapter, the fact that restrictionists were equally concerned with excluding defectives from northwestern Europe, and that inferior races were defined as those with large numbers of defectives, shows that disability was the more fundamental distinction; see *Whiteness of a Different Color: European Immigrants and the Alchemy of Race* (Cambridge, MA: Harvard University Press, 1998), 68–9.

141. On disability and the construction of race, see David R. Roediger, *Seizing Freedom: Slave Emancipation and Liberty for All* (Brooklyn: Verso, 2014), 67–103; Douglas C. Baynton, "Disability and the Justification of Inequality in American History," in *The New Disability History: American Perspectives*, ed. Paul Longmore and Lauri Umansky (New York: New York University Press, 2001), 33–57. On the association between particular nationalities and disabilities, see Anne-Emmanuelle Birn, "Six Seconds per Eyelid: The Medical Inspection of Immigrants at Ellis Island, 1892–1914," *Dynamis* 17 (1997): 281–316.

142. See, for example, Paul K. Longmore and David Goldberger, "The League of the Physically Handicapped and the Great Depression: A Case Study in the New Disability History," *Journal of American History* 87, no. 3 (December 1, 2000): 888–922; Catherine Kudlick, "The Out-

look of *The Problem* and the Problem with the *Outlook*: Two Advocacy Journals Reinvent Blind People in Turn-of-the-Century America," in *The New Disability History: American Perspectives*, ed. Paul K. Longmore and Lauri Umansky (New York: New York University Press, 2001), 187–213; John V. Van Cleve, *A Place of Their Own: Creating the Deaf Community in America* (Washington, DC: Gallaudet University Press, 1989), 87–97.

143. G. K. Chesterton, *Eugenics and Other Evils* (London: Cassell, 1922), 178–79.

144. See Trent, *Inventing the Feeble Mind*, 131–83; Susan M. Schweik, *The Ugly Laws: Disability in Public* (New York: New York University Press, 2009), 63–84, 165–204; Pernick, *The Black Stork*, 41–80.

Chapter Two

1. For a fascinating although very different kind of exploration of the connections between time and disability, see Alison Kafer, *Feminist, Queer, Crip* (Bloomington: Indiana University Press, 2013), 25–68. Portions of chapter 2 appeared in "'These Pushful Days': Time and Disability in the Age of Eugenics," *Health and History* 13, no. 2 (2011): 43–64. Portions of chapters 2, 3, and 4 are reprinted from "Disability and the Justification of Inequality in American History," in Paul Longmore and Lauri Umansky, eds., *The New Disability History: American Perspectives*, New York University Press, 2001, pp. 33–57, with permission from New York University Press © 2001.

2. *The Works of Thomas Jefferson*, ed. Paul Leicester Ford (New York: G. P. Putnam's Sons, 1904–5), 1:89, 1:450, 6:207; *The Writings of Thomas Jefferson: Being His Autobiography, Correspondence, Reports, Messages, Addresses, and Other Writings, Official and Private*, vol. 1, ed. Henry Augustine Washington (Washington, DC: H. W. Derby, 1859), 58, 303. Herman Melville, *Moby-Dick; or, The Whale* (New York: Harper & Brothers, 1851), 171, 391.

3. "Patience and Her Friend" (American Tract Society, 1859), 1, Disability History Museum, http://www.disabilitymuseum.org/dhm/lib/catcard.html?id=56 (accessed May 1, 2015).

4. Penny L. Richards, "'Beside Her Sat Her Idiot Child': Families and Developmental Disability in Mid-Nineteenth-Century America," in *Mental Retardation in America: A Historical Reader*, ed. Steven Noll and James W. Trent (New York: New York University Press, 2004), 68.

5. J. A. Ayres, "An Inquiry into the Extent to Which the Misfortune of Deafness May Be Alleviated," *American Annals of the Deaf and Dumb* 1 (April 1848): 221; Collins Stone, "On the Religious State and Instruction of the Deaf and Dumb," *American Annals of the Deaf and Dumb* 1 (April 1848): 133; Thomas Gallaudet and Thomas Machinator, "Discussion," *Proceedings of the Ninth Convention of American Instructors of the Deaf, 1878* (Columbus, OH, 1879), 131, 133.

6. James Caughey, *Helps to a Life of Holiness and Usefulness, or, Revival Miscellanies: Containing Eleven Revival Sermons and Thoughts on Entire Sanctification* (Boston: James P. Magee, 1852), 403.

7. William Paley, *Natural Theology; or, Evidences of the Existence and Attributes of the Deity, Collected from the Appearances of Nature* (London: W. Mason, Baldwin, 1917), 6–7.

8. Richard Owen, *On the Archetype and Homologies of the Vertebrate Skeleton* (London: Johyn Van Voorst, 1848), 172. Owen attributes these words to John Barclay, *An Inquiry into the Opinions, Ancient and Modern, Concerning Life and Organization* (Edinburgh: Bell & Bradfute, Waugh & Innes, 1822), but while Barclay wrote many similar passages, I was unable to locate this one.

9. Nora Barlow, ed., *The Autobiography of Charles Darwin, 1809–1882* (London: Collins, 1958), 87.

10. James Monteith, *Physical and Intermediate Geography* (New York: A. S. Barnes, 1866), preface, 9, 15–17; Arnold Guyot, *Physical Geography* (New York: Charles Scribner, 1873), 46; John Brocklesby, *Elements of Physical Geography* (Philadelphia: E. H. Butler, 1875), preface, 30–31; Arnold Guyot, *Intermediate Geography* (New York: Scribner, Armstrong, 1867), 7, 9.

11. *An Address, Written by Mr. Clerc: And Read by His Request at a Public Examination of the Pupils in the Connecticut Asylum before the Governour and Both Houses of the Legislature*, May 28, 1818 (Hartford, CT: Hudson, 1818).

12. "Origin of the Handicap," *St. Louis Globe-Democrat*, July 21, 1877. See also *OED Online*, http://www.oed.com/view/Entry/83859?rskey=XjjBBc&result=1, s.v. "handicap" (accessed September 15, 2015); Douglas C. Baynton, "Handicap," in *The Encyclopedia of American Disability History* (New York: Facts on File, 2009).

13. "Southern Republicans," *New York Times*, October 15, 1874. *Galveston Daily News*, March 27, 1897. William Z. Ripley, "Races in the United States," *Atlantic Monthly*, December 1908, 753.

14. See John C. Greene, *Science, Ideology, and World View* (Berkeley and Los Angeles: University of California Press, 1981), 52; Peter J. Bowler, *Evolution: The History of an Idea* (Berkeley and Los Angeles: University of California Press, 1989), 188. Alvar Ellegard's exhaustive review of the British popular press showed that by 1870 the basic idea of evolution was widely accepted in Britain; *Darwin and the General Reader: The Reception of Darwin's Theory of Evolution in the British Periodical Press, 1859–1872* (G teborg, 1958). While in the United States there has been no survey as extensive, Richard Hofstadter's more limited study led him to conclude that, within ten years of publication of *On the Origin of Species*, popular magazines progressed "from hostility to skepticism to gingerly approval and finally to full-blown praise"; Hofstadter, *Social Darwinism in American Thought* (Boston: Beacon Press, 1955), 22. Daniel Levine concluded that "social Darwinism was not so much a conservative doctrine as a universal doctrine. The analogy found a home in America with amazing speed and ubiquity"; Levine, *Jane Addams and the Liberal Tradition* (Madison: University of Wisconsin Press, 1971), 94.

15. Hofstadter, *Social Darwinism in American Thought*, 57.

16. John Brocklesby, *Elements of Physical Geography* (Philadelphia: E. H. Butler, 1868), preface. James Bowen, *Rand-McNally Grammar School Geography* (New York: Rand, McNally, 1901), preface.

17. Jacques W. Redway, Butler's Elementary Geography, (Philadelphia: E. H. Butler, 1888), 5. Alex E. Frye, *Grammar School Geography* (Boston: Ginn, 1902), 2.

18. Scott A. Sandage, *Born Losers: A History of Failure in America* (Cambridge, MA: Harvard University Press, 2005), 11–12. The anthropologist Jean La Fontaine argues that in modern societies "with bureaucratic hierarchies, society is conceived of as an organization of competing individuals" rather than persons valued for their positions in kin and other groups; quoted in Ida Nicolaisen, "Persons and Nonpersons: Disability and Personhood among the Punan Bah of Central Borneo," in *Disability and Culture*, ed. Benedicte Ingstad and Susan Reynolds Whyte (Berkeley and Los Angeles: University of California Press, 1995), 52–53.

19. William Rathbone Greg, "Life at High Pressure," *Contemporary Review* 25 (March 1875): 623–24. *New York Times*, "Life at High Pressure," March 14, 1875. Josiah Strong, *Our Country: Its Possible Future and Its Present Crisis* (New York: Baker & Taylor, 1891), 16, 120, 223.

20. H. G. Wells, "The New Accelerator," *Strand*, December 1901, 622–23. On the effects of the market economy and industrialization on blind workers, see Mary Klages, *Woeful Afflictions: Disability and Sentimentality in Victorian America* (Philadelphia: University of Pennsylvania Press, 1999), 38–46.

21. Samuel Haber, *Efficiency and Uplift: Scientific Management in the Progressive Era* (Chicago: University of Chicago Press, 1973), ix. See also Cecilia Tichi, *Shifting Gears: Technology, Literature, Culture in Modernist America* (Chapel Hill: University of North Carolina Press, 1987), 75–96.

22. Frederick Winslow Taylor, *The Principles of Scientific Management* (New York: Norton, 1967), 8. Michael Knoll, "From Kidd to Dewey: The Origin and Meaning of 'Social Efficiency,'" *Journal of Curriculum Studies* 41, no. 3 (June 2009): 361.

23. Frank W. Booth, "Normal Training for Oral Teachers of the Deaf," *Association Review* 9 (February–April 1907): 206; R. O. Johnson, "Discussion," *Proceedings of the Twentieth Convention of American Instructors of the Deaf, 1914* (Washington, DC, 1915), 186.

24. James W. Trent, *Inventing the Feeble Mind: A History of Mental Retardation in the United States* (Berkeley and Los Angeles: University of California Press, 1994), 60–95; Sarah Frances Rose, "No Right to Be Idle: The Invention of Disability, 1850–1930" (Ph.D. diss., University of Chicago, 2008), 138–42.

25. Edward M. Gallaudet, "Our Profession," *American Annals of the Deaf* 37 (January 1892): 7; Grace C. Green, "The Importance of Physical Training for the Deaf," *Association Review* 9 (February–April 1907): 183. For similar examples, see Henry Smith Williams, "What Shall Be Done with Dependent Children," *North American Review* 164, no. 485 (April 1897): 408 ("handicapped in the struggle for existence"). "The Disabled First," *New York Times*, April 27, 1920 (describing a congressional bill for the training of veterans "whose disabilities are a vocational handicap," in order to "fit them for the battle of life").

26. "The Late Edwin Cowles," *Yenowine's News* (Milwaukee, WI), March 16, 1890. "Handicapped Children: The Care of Those Who Are Physically Disabled," *Rocky Mountain News* (Denver, CO), April 26, 1891.

27. *The Federal Reporter with Key-Number Annotations, Permanent Edition, Cases Argued and Determined in the Circuit Courts of Appeals, Circuit and District Courts, and Commerce Court of the United States, June–July, 1912*, vol. 195 (St. Paul, MN: West, 1912), 129; quoted in Susan Schweik, "Disability and the Normal Body of the (Native) Citizen," *Social Research* 78, no. 2 (Summer 2011): 423.

28. Randolph Bourne, "The Handicapped—By One of Them," *Atlantic Monthly*, September 1911, 320–29; Lillian D. Wald, *The House on Henry Street* (New York: Henry Holt, 1915), 117; Frank Gilbreth and Lillian Gilbreth, *Motion Study for the Handicapped* (London: George Routledge & Sons, 1920), xv.

29. National Archives, RG 85, Records of the INS, entry 7, file no. 49731/2. Perelman arrived at the Port of Boston on December 4, 1905, and was ordered deported on December 6.

30. National Archives, RG 85, Records of the INS, entry 7, file no. 48482. Unger arrived on March 22, 1905, and was ordered deported on March 30.

31. National Archives, Records of the INS, RG 85, Accession 60A600, file 52880/113. Mydland was ordered deported by the board but admitted on appeal when his uncle in Kansas sent an affidavit promising him a job on his farm.

32. Edith Abbott, *Immigration: Select Documents and Case Records* (Chicago: University of Chicago Press, 1924), 251.

33. Norman S. Dike, "Aliens and Crime," in Grant and Davison, *Alien in Our Midst*, 81.

34. Woodbridge N. Ferris, "Governor's Address," *Journal of Psycho-Asthenics* 13 (December 1913): 71.

35. Bleecker Van Wagenen, "Surgical Sterilization As a Eugenic Measure," *Journal of Psycho-Asthenics* 18, no. 4 (June 1914): 187–88.

36. William Roscoe Thayer, "Throwing Away Our Birthright," *North American Review* 215, no. 795 (February 1, 1922): 152.

37. Samuel Gridley Howe, *Report Made to the Legislature of Massachusetts, upon Idiocy* (Boston: Coolidge & Wiley, 1848), 4. See also Allison C. Carey, *On the Margins of Citizenship: Intellectual Disability and Civil Rights in Twentieth-Century America* (Philadelphia: Temple University Press, 2009), 49; Philip M Ferguson, *Abandoned to Their Fate: Social Policy and Practice toward Severely Retarded People in America, 1820–1920* (Philadelphia: Temple University Press, 1994), 16; Peter L. Tyor and Leland Virgil Bell, *Caring for the Retarded in America: A History* (San Francisco: Greenwood Press, 1984), 15, 58; Gerald N. Grob, "The State Mental Hospital in Mid-Nineteenth Century America: A Social Analysis," *American Psychologist* 21, no. 6 (June 1966): 511; James W. Trent, *The Manliest Man: Samuel G. Howe and the Contours of Nineteenth-Century American Reform* (Boston: University of Massachusetts Press, 2012), 152.

38. John Langdon Down, "Observations on an Ethnic Classification of Idiots," *London Hospital Clinical Reports* 3 (1866): 259–62. See also David Wright, "Mongols in Our Midst: John Langdon Down and the Ethnic Classification of Idiocy, 1858–1924," in *Mental Retardation in America: A Historical Reader*, ed. Steven Noll and James W. Trent (New York: New York University Press, 2004), 136–86; David Wright, *Down's Syndrome: The History of a Disability* (New York: Oxford University Press, 2011), 48–50.

39. C. P. Knight, "The Detection of the Mentally Defective among Immigrants," *Journal of the American Medical Association* 60 (1913): 106–7.

40. Daniel J. Kevles, *In the Name of Eugenics: Genetics and the Uses of Human Heredity* (Berkeley and Los Angeles: University of California Press, 1985), 160. Francis Graham Crookshank, *The Mongol in Our Midst: A Study of Man and His Three Faces* (New York: E. P. Dutton, 1924), 4–7.

41. Steven A. Gelb, "The Problem of Typological Thinking in Mental Retardation," *Mental Retardation* 35 (1997): 448–57.

42. Charles Darwin, *The Descent of Man, and Selection in Relation to Sex* (London: John Murray, 1874), 36.

43. John Peter Lesley, *Man's Origin And Destiny: Sketched From the Platform of the Physical Sciences* (Philadelphia: J. B. Lippincott, 1868), 120.

44. George E. Vincent, "A Retarded Frontier," *American Journal of Sociology* 4, no. 1 (July 1898): 1–20.

45. George W. Twitmyer, "Clinical Studies of Retarded Children," *Psychological Clinic* 1, no. 4 (June 15, 1907): 97–98, 102. See also J. David Smith and Gretchen Smallwood, "From Whence [*sic*] Came Mental Retardation? Asking Why While Saying Goodbye," *Intellectual and Developmental Disabilities* 45, no. 2 (April 2007): 132–35.

46. David B. Tyack, *The One Best System: A History of American Urban Education* (Cambridge, MA: Harvard University Press, 1974), 33; Richards, "'Beside Her Sat Her Idiot Child,'" 67; Lightner Witmer, "The Study and Treatment of Retardation: A Field of Applied Psychology," *Psychological Bulletin* 6, no. 4 (April 15, 1900): 123; Julia Richman, "What Can be Done in a Graded School for the Backward Child," *Charities* 13 (November 5, 1904): 129–31.

47. C. H. Henninger, "The Feeble-Minded Outside the Institution and Their Relation to Society," *Journal of Psycho-Asthenics* 16, no. 4 (June 1912): 154.

48. Henry H. Goddard, "Suggestions for a Prognostical Classification of Mental Defectives," *Journal of Psycho-Asthenics* 14 (1909–10): 53.

49. Seymour Bernard Sarason and John Doris, *Educational Handicap, Public Policy, and Social History* (New York: Free Press, 1979).

50. U.S. Department of Commerce, Bureau of the Census, *Insane and Feebleminded in Institutions, 1910* (Washington, DC: Government Printing Office, 1914), 183.

51. Oliver P. Cornman, "The Retardation of the Pupils of Five City School Systems," *Psychological Clinic* 1, no. 9 (February 15, 1908), 254–255. See also Smith and Smallwood, "From Whence Came Mental Retardation?"

52. Felix Arnold, "Politics, Efficiency, and Retardation," *Psychological Clinic* 7, no. 1 (March 15, 1913), 36.

53. Twitmyer, "Clinical Studies of Retarded Children," 97–98, 102.

54. R. H. Sylvester, "Clinical Psychology Adversely Criticized," *Psychological Clinic* 7, no. 7 (December 15, 1913): 182, 185, 188. C. S. Berry, "Intelligence Quotients of Mentally Retarded School Children," *School and Society* 17 (1923):723–29. Anna Johnson, "Retarded Sixth Grade Pupils," *Psychological Clinic* 7, no. 6 (November 15, 1913): 163.

55. *Proceedings of the Association of Medical Officers of American Institutions for Idiotic and Feebleminded Persons* (Philadelphia: J. B. Lippincott, 1876), 32.

56. Stanley D. Porteus, "Individual and Racial Retardation," *Training School Bulletin* 20 (October 1923): 84–85.

57. Herbert M. Kliebard, *The Struggle for the American Curriculum, 1893–1958* (New York: Routledge, 2004), 78, 88.

58. Elisabeth Antoinette Irwin, Truancy: A Study of the Mental, Physical and Social Factors of the Problem of Non-Attendance at School (New York: Public Education Association of the City of New York, 1915), 61, 66. See also Richard James Arthur Berry and Stanley David Porteus, Intelligence and Social Valuation: A Practical Method for the Diagnosis of Mental Deficiency and Other Forms of Social Inefficiency (Vineland, NJ: Publications of the Training School at Vineland New Jersey, 1920).

59. Irwin, *Truancy*, 14, 23, 27, 61, 64–65.

60. Thomas C. Leonard, "'More Merciful and Not Less Effective': Eugenics and American Economics in the Progressive Era," *History of Political Economy* 35, no. 4 (Winter 2003): 702.

61. Henry R. Seager, "The Minimum Wage as Part of a Program for Social Reform," *Annals of the American Academy of Political and Social Science* 48 (July 1, 1913): 910. See Leonard, "'More Merciful and Not Less Effective,'" for examples of other economists who held similar views.

62. William Trufant Foster, *The Social Emergency: Studies in Sex Hygiene and Morals* (Boston: Houghton Mifflin, 1914), 18.

63. "Commission for Moral Efficiency," *Pittsburgh Gazette Times*, May 26, 1912; Pittsburgh Morals Efficiency Commission, *Report and Recommendations of Morals Efficiency Commission, 1913* (Pittsburgh: Pittsburgh Printing, 1913); Economic Club of New York, *Yearbook of the Economic Club of New York: Containing the Addresses at the Four Meetings of the Season* (1913), 71–72.

64. Emil G. Hirsch, "Religious Education and Moral Efficiency," *Religious Education* 4, no. 2 (1909): 129–30.

65. Trent, *Inventing the Feeble Mind*, 167–69; Ian Robert Dowbiggin, *Keeping America Sane: Psychiatry and Eugenics in the United States and Canada, 1880–1940* (Ithaca, NY: Cornell University Press, 1997), 191–93.

66. Gilbert L. Brown, "Intelligence as Related to Nationality," *Journal of Educational Research* 5, no. 4 (April 1922): 324–27. See also Rudolf Pintner, "Comparison of American and Foreign Children on Intelligence Tests," *Journal of Educational Psychology* 14 (1923): 292–95.

67. Howard A. Knox, "The Moron and the Study of Alien Defectives," *Journal of the American Medical Association* 60 (1913): 105–6.

68. Howard Knox, *Manual of the Mental Examination of Aliens*, Miscellaneous Publication, United States Public Health Service (Washington, DC: Government Printing Office, 1918). See also E. H. Mullan, "Mental Examination of Immigrants: Administration and Line Inspection at Ellis Island," *Public Health Reports* 32, no. 20 (May 18, 1917): 746. Anti-Mexican-immigrant rhetoric in the 1920s emphasized inherent intellectual disability; see Alexandra Minna Stern, *Eugenic Nation: Faults and Frontiers of Better Breeding in Modern America* (Berkeley and Los Angeles: University of California Press, 2005), 91, 97. On "passing" and intellectual disability, see Allison C. Carey, "The Sociopolitical Contexts of Passing and Intellectual Disability," in *Disability and Passing: Blurring the Lines of Identity*, ed. Jeffrey A. Brune and Daniel J. Wilson (Philadelphia: Temple University Press, 2013), 142–66.

69. Letter from Victor Safford to the Commissioner, May 16, 1906, Immigration Restriction League Records, US 10583.9.8—US 10587.43, box 4, folder Safford, M. Victor—"Definitions of Various Medical Terms Used in Medical Certificates, 1906," Houghton Library, Harvard University.

70. National Archives, Records of the INS, RG 85, Accession 60A600, file 52880/113.

71. National Archives, Records of the INS, RG 85, Accession 60A600, file 52,242–11.

72. Georges Canguilhem, *The Normal and the Pathological* (New York: Zone Books, 1989), 39–64, 125; Honoré de Balzac, *Eugénie Grandet* (Oxford: Oxford University Press, 2009), 188; Ian Hacking, *The Taming of Chance* (Cambridge: Cambridge University Press, 1990), 160–66.

73. "Bowen's Political Economy," *North American Review* 82, no. 171 (April 1856): 538–39.

74. James Russell Lowell, writing under the pseudonym, Markiss O' Big Boosy, "Birdofredum Sawin, Esq., to Mr. Hosea Bigelow," *Atlantic Monthly*, March 1862, 389.

75. Hacking, *The Taming of Chance*, 160–66. See also Baynton, *Forbidden Signs*, chapters 5–6.

76. Edmund Burke, *Reflections on the Revolution in France*, ed. J. G. A. Pocock (Cambridge, MA: Hackett, 1987), 60, 168, 171, 188, 190, 203, 216. Thomas Paine, *Rights of Man* (New York: Dover, 1999), 42.

77. Francois Ewald, "Norms, Discipline, and the Law," *Representations* 30 (Spring 1990): 146, 149–50, 154; Lennard J. Davis, *Enforcing Normalcy: Disability, Deafness, and the Body* (New York: Verso, 1995); Baynton, *Forbidden Signs*, chapters 5 and 6.

78. "The Progress of Southern Opinion," *Provincial Freeman* [reprint], November 25, 1856, http://www.accessible.com (accessed September 5, 2011).

79. Francis Galton, "Typical Laws of Heredity," *Proceedings of the Royal Institution*, vol. 8, no. 66 (February 9, 1877): 297.

80. For example, late nineteenth-century educators began using "normal child" as the counterpart to "deaf child," instead of the "hearing" versus "deaf" of previous generations. "Normal" can appear to refer to an average, since the average person is hearing, but it does not exclude those with superior hearing. It denotes those above a certain standard not the average.

81. George Cotkin, *Reluctant Modernism: American Thought and Culture* (New York: Twayne, 1992), xii, 3, 20–24.

82. Arnold Guyot, *Physical Geography* [1873] (New York: American Book, 1885), 114–18. John H. Van Evrie, *White Supremacy and Negro Subordination, or Negroes a Subordinate Race* (New York: Van Evrie, Horton, 1868), 199, chapter 15 passim.

83. James W. Trent, Jr., "Defectives at the World's Fair: Constructing Disability in 1904," *Remedial and Special Education* 19 (July–August 1998): 201–11.

84. Gardiner G. Hubbard, "Proceedings of the American [Social] Science Association," *National Deaf Mute Gazette* 2 (January 1868): 5.

85. J. D. Kirkhuff, "Superiority of the Oral Method," *Silent Educator* 3 (January 1892): 139.

86. J. C. Gordon, "Dr. Gordon's Report," *Association Review* 1 (December 1899): 206.

87. Susanna E. Hull, "Do Persons Born Deaf Differ Mentally from Others Who Have the Power of Hearing?" *American Annals of the Deaf* 22 (October 1877): 236.

88. Emma Garrett, "A Plea That the Deaf 'Mutes' of America May Be Taught to Use Their Voices," *American Annals of the Deaf* 28 (January 1883): 18.

89. Lewis Dudley, "Address of Mr. Dudley in 1880," *Fifteenth Annual Report of the Clarke Institution for Deaf Mutes* (Northampton, MA: Steam Press of Gazette Printing, 1882), 7.

90. Mary McCowen, "How Best to Secure Intelligent Speech for Deaf Children," *Association Review* 9 (February–April 1907): 258–59.

91. Edward Seguin, "Origin of Method of Treatment and Training of Idiots," *American Journal of Education* 2 (1856): 147.

92. Quoted in Trent, *Inventing the Feeble Mind*, 83.

93. Mary Garrett and Emma Garrett, "The Possibilities of the Oral Method for the Deaf and the Next Steps Leading towards Its Perfection," *Silent Educator* 3 (January 1892): 65; Jennie L. Cobb, "Schoolroom Efficiency," *American Annals of the Deaf* 58 (May 1913): 208; *Scientific American*, June 8, 1907, 474; A. L. E. Crouter, "The Possibilities of Oral Methods in the Instruction of the Deaf," *Proceedings of the Nineteenth Convention of American Instructors of the Deaf, 1911* (Washington, DC, 1912), 139–41.

94. "Eugenicists Hear of Real 'Missing Link,'" *New York Times*, January 20, 1914.

95. Ibid.

96. William Shakespeare, *Venus and Adonis* (Stratford-upon-Avon: Richard Field, 1595).

97. Words and music by Frank Wilder, 1863; http://en.wikipedia.org/wiki/The_Invalid_Corps. Quoted in David R. *Roediger, Seizing Freedom: Slave Emancipation and Liberty for All* (Brooklyn: Verso, 2014), 75.

98. Mark Twain, *Life on the Mississippi* (1883; New York: Bantam Dell, 2007), 288.

99. Martin W. Barr, "Mental Defectives and the Social Welfare," *Popular Science Monthly*, April 1899, 747.

100. See Alexandra Minna Stern, *Eugenic Nation: Faults and Frontiers of Better Breeding in Modern America* (Berkeley and Los Angeles: University of California Press, 2005), 13–14; Nicole Hahn Rafter, *Creating Born Criminals* (Urbana: University of Illinois Press, 1997), 36–39.

101. William House, "Constitutional Psychopathic Inferiority," *California State Journal of Medicine* 21 (January 1923): 26–29; Alfred C. Reed, "Immigration and the Public Health," *Popular Science Monthly*, October 1913, 313–38.

102. Letter from Victor Safford to the Commissioner, May 16, 1906, Immigration Restriction League Records, US 10583.9.8—US 10587.43, box 4, folder Safford, M. Victor—"Definitions of Various Medical Terms Used in Medical Certificates, 1906," Houghton Library, Harvard University.

103. Thomas Darlington, "The Medico-Economic Aspect of the Immigration Problem," *North American Review* 183, no. 605 (December 1906): 1269.

104. Letter, June 10, 1911, Immigration Restriction League Records, US10583.9.8—US10587.43, Box 2 "Correspondence: Fe—Q", folder Fisk, Arthur L., Houghton Library, Harvard University.

105. Robert DeCourcy Ward, "Our Immigration Laws from the Viewpoint of National Eugenics," *National Geographic Magazine*, January 1912, 40. On the significance of degeneracy in the social policy of the era, see Nicole Rafter, "The Criminalization of Mental Retardation," in *Mental Retardation in America: A Historical Reader*, ed. Steven Noll and James W. Trent (New York: New York University Press, 2004), 232–57.

106. Irving Fisher, "Impending Problems of Eugenics," *Scientific Monthly* 13, no. 3 (September 1921): 215, 229.

107. Edward A. Ross, "The Causes of Race Superiority," *Annals of the American Academy of Political and Social Science* 18 (July 1, 1901): 88.

108. Theodore Roosevelt, "On American Motherhood," an address to the National Congress of Mothers, March 13, 1905.

109. Charles B. Davenport, "Influence of Heredity on Human Society," *Annals of the American Academy of Political and Social Science* 34 (1909): 20–21.

110. George Frank Lydston and Eugene S. Talbot, "Studies of Criminals," *Alienist and Neurologist* 12 (1891): 557; "The Study of the Criminal," *Boston Medical and Surgical Journal* 125 (July–December 1891): 579. See also Steven A. Gelb, "'Not Simply Bad and Incorrigible': Science, Morality, and Intellectual Deficiency," *History of Education Quarterly* 29, no. 3 (Autumn 1989): 359–79.

111. Goddard, quoted in Rafter, *Creating Born Criminals*, 137–39. See also Henry Martyn Boies, *Prisoners and Paupers: A Study of the Abnormal Increase of Criminals, and the Public Burden of Pauperism in the United States* (New York: Putnam, 1893), 172.

112. Martin Barr, *Mental Defectives: Their History, Treatment, and Training* (Philadelphia: P. Blakiston's, 1904), 100–101.

113. William Williams, "Immigration and Insanity," in *Proceedings of the Mental Hygiene Conference and Exhibit, New York, 1912* (New York: Committee on Mental Hygiene of the State Charities Aid Association, 1912), 180. See also letter from Williams to the Commissioner General, March 31, 1913, National Archives, Records of the INS, RG 85, entry 9, file 51490/19, p. 7.

114. "Putting Our Immigrants through the Sieve: Government Stands as 'Doctor of Eugenics' at Portals of Nation," *Portland Oregonian*, September 28, 1913.

115. Jack London, *The Sea-Wolf*, 2nd ed. (New York: Bantam Classics, 1984), 78.

116. Edgar Rice Burroughs, *The Oakdale Affair*, 2008, http://www.gutenberg.org/ebooks/363 (accessed May 1, 2014).

117. Rosemary Jann, "Sherlock Holmes Codes the Social Body," *ELH* 57, no. 3 (October 1, 1990): 685–708.

118. Daniel Pick, "'Terrors of the Night': Dracula and 'Degeneration' in the Late Nineteenth Century," *Critical Quarterly* 30, no. 4 (December 1, 1988): 80.

119. Francis A. Walker, "Restriction of Immigration," *Atlantic Monthly*, June 1896, 828.

120. Charles B. Davenport, "Influence of Heredity on Human Society," *Annals of the American Academy of Political and Social Science* 34 (1909): 16–21.

121. Allan J. McLaughlin, "The American's Distrust of the Immigrant," *Popular Science Monthly*, January 1903, 231.

122. Davenport, "Influence of Heredity," 17–18; G. A. Doren, *Our Defective Classes: How to Care for Them and Prevent Their Increase* (1899; repr. Columbus, OH: Westbote, 1902), 4. On metaphors of contagion applied to immigration, see also Amy L. Fairchild, *Science at the Borders: Immigrant Medical Inspection and the Shaping of the Modern Industrial Labor Force* (Baltimore, MD: Johns Hopkins University Press, 2003), 42–43.

123. It is not that genetic defect has been ignored, but rather that historians of medical inspection, for the most part, have done little to explore its significance. Alan Kraut's otherwise valuable *Silent Travelers: Germs, Genes, and the "Immigrant Menace"* is one such example; as Martin Pernick pointed out, "Despite his subtitle, Kraut concentrates on infectious diseases, not eugenics." Martin S. Pernick, "Eugenics and Public Health in American History," *American Journal of Public Health* 87, no. 11 (November 1997): 1771 n. 14.

124. Howard Markel and Alexandra Minna Stern, "Which Face? Whose Nation? Immigration, Public Health, and the Construction of Disease at America's Ports and Borders, 1891–1928," *American Behavioral Scientist* 42, no. 9 (June 1, 1999): 1319. See also Howard Markel and Alexandra Minna Stern, "The Foreignness of Germs: The Persistent Association of Immigrants and Disease in American Society," *Milbank Quarterly* 80, no. 4 (January 1, 2002): 763–64.

125. "The Invasion of the Unfit," *Medical Record* 82, no. 24 (December 14, 1912): 1080.

126. Alfred C. Reed, "Immigration and the Public Health," *Popular Science Monthly*, October 1913, 317, 325.

127. Pernick, "Eugenics and Public Health," 1769.

128. National Archives, Records of the INS, RG 85, Accession 60A600, file 51449/24.

129. Davenport, "Influence of Heredity," 18–19.

130. Ernest Bicknell, "Notes and Abstracts: Feeblemindedness as an Inheritance," *American Journal of Sociology* 2, no. 4 (January 1897): 627.

131. National Archives, Records of the INS, RG 85, Accession 60A600, file 51970–10. If an immigrant experienced psychiatric problems up to five years after arrival, a diagnosis of "constitutional psychopathic inferiority" was often used to claim that the disability existed upon arrival and thereby justified deportation. When such findings faced legal challenge, courts typically deferred to the authority of immigration officials; see for example: *United States ex rel. Brugnoli v. Tod, Commissioner of Immigration*, 300 F. 913 (District Court, S.D., New York, July 6, 1923); *United States ex rel. Haft v. Tod, Commissioner of Immigration*, 300 F. 918 (Circuit Court of Appeals, 2nd Cir., April 28, 1924); *United States ex rel. Mandel v. Day, Commissioner of Immigration*, 19 F.2d 520 (District Court, E.D., New York, April 14, 1927); *United States ex rel. Paolantonio v. Day, Commissioner of Immigration*, 22 F.2d 914 (Circuit Court of Appeals, 2nd Cir., December 5, 1927); *United States ex rel. Powlowec v. Day, Commissioner of Immigration*, 33 F.2d 267 (Circuit Court of Appeals, 2nd Cir., June 10, 1929).

Chapter Three

1. National Archives, Records of the INS, RG 85, Accession 60A600, file 53,700/974. Portions of chapter 3 are reprinted by permission of the publisher from "The Undesirability of Admitting Deaf Mutes': American Immigration Policy and Deaf Immigrants, 1882–1924," in *Sign Language Studies*, Vol. 6, No. 4, by Douglas C. Baynton (Washington, DC: Gallaudet University Press, 2006): 391–415. Copyright 2006 by Gallaudet University.

2. On the uses of the dichotomy of independence and dependence in liberal thought and American law, see Allison C. Carey, *On the Margins of Citizenship: Intellectual Disability and Civil Rights in Twentieth-Century America* (Philadelphia: Temple University Press, 2009), 5–7, 16–20, 42–45.

3. Harry Best, *The Deaf: Their Position in Society and the Provision for Their Education in the United States* (New York: Thomas Y. Crowell, 1914), 78–80, 90.

4. Harry Best, *Blindness and the Blind in the United States* (New York: Macmillan, 1934), 80–82.

5. National Archives, Records of the INS, RG 85, entry 7, file 13,583.

6. See Paul Longmore and David Goldberger, "The League of the Physically Handicapped and the Great Depression: A Case Study in the New Disability History," *Journal of American History* 87 (December 2000): 888–922.

7. The historian Deirdre Moloney suggests that dependence was also used to exclude women suspected of being prostitutes because it was easier to demonstrate. Deirdre M. Moloney,

"Women, Sexual Morality, and Economic Dependency in Early U.S. Deportation Policy," *Journal of Women's History* 18, no. 2 (Summer 2006): 98. See also Barbara Young Welke, *Law and the Borders of Belonging in the Long Nineteenth Century United States* (Cambridge: Cambridge University Press, 2010), 75.

8. See Jeanne D. Petit, *The Men and Women We Want: Gender, Race, and the Progressive Era Literacy Test Debate* (Rochester, NY: University of Rochester Press, 2010), 44, 88, 92. Donna Gabaccia, *From the Other Side: Women, Gender, and Immigration Life in the U.S., 1820–1990* (Bloomington: Indiana University Press, 1994), 37. Martha Gardner, *The Qualities of a Citizen: Women, Immigration, and Citizenship, 1870–1965* (Princeton, NJ: Princeton University Press, 2005), 87–99.

9. Edward Clarke, *Sex in Education; or, A Fair Chance for Girls* (1873; New York: Arno Press, 1972), 18, 22, 62.

10. William Warren Potter, "How Should Girls Be Educated? A Public Health Problem for Mothers, Educators, and Physicians," *Transactions of the Medical Society of the State of New York* (1891): 48.

11. Arthur Lapthorn Smith, "Higher Education of Women and Race Suicide," *Popular Science Monthly*, March 1905, 467–69, 473.

12. Almroth Edward Wright, *The Unexpurgated Case against Woman Suffrage* (New York: Paul B. Hoeber, 1913), 41, 79, 82, 88, 93, 98, 168–9. The *New York Times* announced its publication in London, printed excerpts in its *Sunday Magazine*, and announced its publication in the United States; *New York Times*, October 1, 1913; *New York Times*, October 19, 1913; *New York Times*, October 26, 1913.

13. "The Unexpurgated Case against Woman Suffrage," *Saturday Review of Politics, Literature, Science and Art*, October 12, 1913, 14.

14. Charles L. Dana, "Suffrage a Cult of Self and Sex: The Average Zealot has the Mental Age of Eleven and through a Cranny Sees the Dazzling Illusion of a New Heaven," *New York Times*, June 27, 1915.

15. Grace Duffield Goodwin, *Anti-Suffrage: Ten Good Reasons* (New York: Duffield, 1913), 91–92.

16. Clarke, *Sex in Education*, 12.

17. Lisa Tickner, *The Spectacle of Women: Imagery of the Suffrage Campaign, 1907–14* (Chicago: University of Chicago Press, 1988), illustration IV; Alice Sheppard, *Cartooning for Suffrage* (Albuquerque: University of New Mexico Press, 1994), 30; Elizabeth Cady Stanton, Susan B. Anthony, and Matilda Joslyn Gage, eds., *History of Woman Suffrage*, vol. 2 (New York: Arno Press, 1969), 288; Elizabeth Cady Stanton, "Address to the National Woman Convention, Washington, D.C., January 19, 1869," in *The Concise History of Woman Suffrage: Selections from "History of Woman Suffrage,"* ed. Mari Jo Buhle and Paul Buhle (Urbana: University of Illinois Press, 1978), 256.

18. Lois N. Magner, "Darwinism and the Woman Question: The Evolving Views of Charlotte Perkins Gilman," in *Critical Essays on Charlotte Perkins Gilman*, ed. Joanne Karpinski (New York: G. K. Hall, 1992), 119–20. Nancy Woloch, *Women and the American Experience*, vol. 1, *To 1920* (New York: McGraw-Hill, 1994), 339. Aileen S. Kraditor, *The Ideas of the Woman Suffrage Movement* (New York: W. W. Norton, 1981), 20. See also Anne Digby, "Woman's Biological Straitjacket," in *Sexuality and Subordination: Interdisciplinary Studies of Gender in the Nineteenth Century*, ed. Susan Mendas and Jane Randall (New York: Routledge, 1989), 192–220; Martha H. Verbrugge, *Able-Bodied Womanhood: Personal Health and Social Change in Nineteenth-Century Boston* (Oxford and New York: Oxford University Press, 1988), 120–22; Jane Jerome Camhi,

Women against Women: American Anti-Suffragism, 1880–1920 (New York: Carlson, 1994), 18; Mara Mayor, "Fears and Fantasies of the Anti-Suffragists," *Connecticut Review* 7 (April 1974): 67.

19. National Archives, Records of the INS, RG 85, entry 7, file 50004–2.

20. U.S. Immigration Service, *Annual Report of the Superintendent of Immigration* (Washington, DC: Government Printing Office, 1894), 12–13.

21. Gardner, *The Qualities of a Citizen*, 88.

22. The policy applied also to families that were willing to be separated: in 1911, on her way to join her husband, Angela Esposito wished to continue on her way while her son, certified as an "idiot," was returned to Italy. Officials sent them back together. Gardner, *The Qualities of a Citizen*, 26.

23. National Archives, Records of the INS, RG 85, Accession 60A600, file 53,595/462.

24. National Archives, Records of the INS, RG 85, Accession 60A600, file 53,550/580.

25. National Archives, Records of the INS, RG 85, Accession 60A600, file 53550/269. Referred to in the documents as the "case of Pesche Straczynsky, child and nephew." Schie Budwicky was traveling with his aunt to join his parents, who were already in the United States.

26. National Archives, Records of the INS, RG 85, Accession 60A600, file 54541/93.

27. National Archives, Records of the INS, RG 85, entry 7, file 49362/4.

28. National Archives, Records of the INS, RG 85, Accession 60A600, file 53,612/336.

29. National Archives, Records of the INS, RG 85, Accession 60A600, file 53,612/88. See also the case of a deaf Polish immigrant, Hirsch B. Wanderman, a journeyman tailor turned back as a likely public charge, despite assurances from his prosperous brother and uncle, both citizens in the clothing business, to assist him; *United States ex rel. Engel v. Tod*, 294 F. 820 (2nd Cir., December 3, 1923).

30. National Archives, Records of the INS, RG 85, entry 7, file 47552.

31. National Archives, Records of the INS, RG 85, Accession 60A600, file 53,470/53.

32. National Archives, Records of the INS, RG 85, entry 7, file 49833–2. On the exclusion and deportation of women deemed immoral or likely to become a public charge, see Gardner, *The Qualities of a Citizen*, 80–99.

33. See Douglas C. Baynton, "Defectives in the Land: Disability and American Immigration Policy, 1882–1924," *Journal of American Ethnic History* (Spring 2005): 37–38.

34. National Archives, Records of the INS, RG 85, Accession 60A600, file 53,385/146.

35. Vincent J. Cannato, *American Passage: The History of Ellis Island* (New York: Harper, 2009), 185, 189.

36. National Archives, Records of the INS, RG 85, entry 7, file 49,968/3.

37. National Archives, Records of the INS, RG 85, entry 7, file 49362/4. Kremen arrived at Ellis Island on August 29, 1905, and was ordered deported on September 7.

38. National Archives, Records of the INS, RG 85, entry 7, file 48482. Unger arrived on March 22, 1905, and was ordered deported on March 30.

39. National Archives, Records of the INS, RG 85, entry 7, file 49824/4. Feinberg arrived on December 27, 1905 and was ordered admitted on January 2, 1906, on appeal after a bond was provided by her sister and brother-in-law.

40. National Archives, Records of the INS, RG 85, entry 7, file 49731/2. Perelman arrived at the Port of Boston on December 4, 1905, and was ordered deported on December 6.

41. National Archives, Records of the INS, RG 85, entry 7, file 48310. Potts arrived on February 24, 1905, and was admitted on March 2 on appeal after a bond was provided by her parents.

42. National Archives, Records of the INS, RG 85, Accession 60A600, file 54577/187. Martinez

arrived at El Paso on April 14, 1919, and was admitted on June 23 on appeal after a bond was provided by his parents.

43. National Archives, Records of the INS, RG 85, Accession 60A600, file 52880/314. Jacobsen arrived on April 1, 1910 and was admitted on April 9 on appeal after her sister and brother furnished a bond and the commissioner determined that she had approximately $2,000 in a bank in Norway.

44. National Archives, Records of the INS, RG 85, entry 7, file 54577/187. See also "Aged Missionaries Barred," *New York Times*, May 3, 1905.

45. Letter from Randolph Kingsley to Frederic C. Howe, March 24, 1916, National Archives, Records of the INS, RG 85, Accession 60A600, file 54050/726.

46. Letter from Randolph Kingsley to Frederic C. Howe, May 10, 1916, National Archives, Records of the INS, RG 85, Accession 60A600, file 54050/726.

47. Letter from George Kessler to William Wilson, May 12, 1916, National Archives, Records of the INS, RG 85, Accession 60A600, file 54050/726.

48. Letters from Frederic C. Howe to the commissioner general, May 12, 1916, and from Alfred Hampton, Assistant Commissioner General, to Commissioner of Immigration, Ellis Island, May 13, 1916, National Archives, Records of the INS, RG 85, Accession 60A600, file 54050/726.

49. National Archives, Records of the INS, RG 85, Accession 60A600, file 54,670/331. Some of the Immigration Bureau records misspell his name as "Hazilstad."

50. National Archives, Records of the INS, RG 85, Accession 60A600, file 54,670/331.

51. Letter from Senator Knute Nelson to Anthony Caminetti, Commissioner General of Immigration, September 19, 1919, National Archives, Records of the INS, RG 85, Accession 60A600, file 54,670/331.

52. Letter from H. Bryn to Robert Lansing, Secretary of State, September 19, 1919, National Archives, Records of the INS, RG 85, Accession 60A600, file 54,670/331.

53. Letter from James Creese Jr., American Scandinavian Foundation, to Anthony Caminetti, Commissioner General of Immigration, September 22, 1919, National Archives, Records of the INS, RG 85, Accession 60A600, file 54,670/331.

54. September 24, 1919, National Archives, Records of the INS, RG 85, Accession 60A600, file 54,670/331.

55. September 26, 1919, National Archives, Records of the INS, RG 85, Accession 60A600, file 54,670/331.

56. September 26, 1919, National Archives, Records of the INS, RG 85, Accession 60A600, file 54,670/331.

57. National Archives, Records of the INS, RG 85, Accession 60A600, file 54,670/331.

58. National Archives, Records of the INS, RG 85, Accession 60A600, file 52880/171. See also Howard Markel, *When Germs Travel: Six Major Epidemics That Have Invaded America since 1900 and the Fears They Have Unleashed* (New York: Pantheon Books, 2004), 35–36.

59. National Archives, Records of the INS, RG 85, Accession 60A600, file 54766/634.

60. National Archives, Records of the INS, RG 85, Accession 60A600, file 53,700/974.

Chapter Four

1. Martin Pernick, *The Black Stork: Eugenics and the Death of "Defective" Babies in American Medicine and Motion Pictures since 1915* (Oxford and New York: Oxford University Press, 1996), 64.

2. Bruce Clayton, *Forgotten Prophet: The Life of Randolph Bourne* (Columbia: University of Missouri Press, 1998), 220.

3. Letter, April 29, 1905, National Archives, Record Group 85, Records of the Immigration and Naturalization Service, entry 7, file no. 48,599/4.

4. See, for example, S. Kay Toombs, "Disability and the Self," in *Changing the Self: Philosophies, Techniques, and Experiences*, ed. Thomas M. Brinthaupt and Richard P. Lipka (New York: SUNY Press, 1994), 344; Jenny Morris, *Pride against Prejudice* (Philadelphia: New Society, 1991), 192.

5. Frances Cooke Macgregor, "Some Psychosocial Problems Associated with Facial Deformities," *American Sociological Review* 16 (1951): 629–30. Marsha R. Peterson, "Yes, There Is a Duty to Accommodate Someone Regarded as Disabled under the ADA," *Nevada Law Journal* 7 (2007): 616–17.

6. Rosemarie Garland-Thomson, *Extraordinary Bodies: Figuring Physical Disability in American Culture and Literature* (New York: Columbia University Press, 1997), 12.

7. Erving Goffman, *Stigma: Notes on the Management of Spoiled Identity* (Prentice Hall, NJ: Simon & Schuster, 1963). See also Lerita Coleman Brown, "Stigma: An Enigma Demystified," in *The Disability Studies Reader*, ed. Lennard J. Davis (New York: Routledge, 2010), 184.

8. See, for example, R. E. Kleck, "Emotional Arousal in Interactions with Stigmatized Persons," *Psychological Reports* 19 (1966): 12–26; R. E. Kleck, H. Ono, and A. H. Hastorf, "The Effect of Physical Deviance upon face-to-face Interaction," *Human Relations* 19 (1966): 425–36; R. E. Kleck, "Physical Stigma and Nonverbal Cues Emitted in Face-to-Face Interaction," *Human Relations* 21 (1968): 19–28; Melvin L. Snyder, Robert E. Kleck, Angelo Strenta, and Steven J. Mentzer, "Avoidance of the Handicapped: An Attributional Ambiguity Analysis," *Journal of Personality and Social Psychology* 37, no. 12 (1979): 2297–2306.

9. Harlan Hahn, "Antidiscrimination Laws and Social Research on Disability: The Minority Group Perspective," *Behavioral Sciences and the Law* 14, no. 1 (Winter 1996): 54. See also Lennard J. Davis, *Enforcing Normalcy: Disability, Deafness, and the Body* (New York: Verso, 1995), 11–12.

10. Richard J. Stevenson, Trevor I. Case, and Megan J. Oaten, "Proactive Strategies to Avoid Infectious Disease," *Philosophical Transactions of the Royal Society B: Biological Sciences* 366, no. 1583 (October 31, 2011): 3362. See also Stephen Ryan, Megan Oaten, Richard J. Stevenson, and Trevor I. Case, "Facial Disfigurement Is Treated like an Infectious Disease," Evolution and Human Behavior 33, no. 6 (November 2012): 639–46.

11. Megan Oaten, Richard J. Stevenson, and Trevor I. Case, "Disease Avoidance as a Functional Basis for Stigmatization," *Philosophical Transactions of the Royal Society B: Biological Sciences* 366, no. 1583 (October 31, 2011): 3435.

12. Valerie Curtis, "Why Disgust Matters," *Philosophical Transactions of the Royal Society B: Biological Sciences* 366, no. 1583 (October 31, 2011): 3484. See also J. H. Park, J. Faulkner, and M. Schaller, "Evolved Disease-Avoidance Processes and Contemporary Anti-Social Behavior: Prejudicial Attitudes and Avoidance of People with Physical Disabilities," *Journal of Nonverbal Behavior* 27 (2003): 65–87.

13. Martha C. Nussbaum, *From Disgust to Humanity: Sexual Orientation and Constitutional Law* (Oxford: Oxford University Press, 2010), 15–17.

14. Martha C. Nussbaum, *Hiding from Humanity: Disgust, Shame, and the Law* (Princeton, NJ: Princeton University Press, 2004), 14 and chapter 2.

15. Alasdair C. MacIntyre, *Dependent Rational Animals: Why Human Beings Need the Virtues* (Chicago: Carus, 1999), 4.

16. Alfred C. Reed, "The Medical Side of Immigration," *Popular Science Monthly*, April 1912,

384–90; Allan J. McLaughlin, "Immigration and the Public Health," in *Public Health Papers and Reports* (Columbus, OH: Press of Fred J. Heer, 1904), 224–31.

17. George Lydston, *The Diseases of Society and Degeneracy* (Philadelphia: J. B. Lippincott, 1908), 129, 132–33.

18. Quoted in Daniel E. Bender, *American Abyss: Savagery and Civilization in the Age of Industry* (Ithaca, NY: Cornell University Press, 2009), 202.

19. James John Davis, *Selective Immigration* (St. Paul, MN: Scott-Mitchell, 1925), 112.

20. Elizabeth F. Frazer, "Our Foreign Cities: New York," *Saturday Evening Post,* June 16, 1923, 6–7. For other examples and an illuminating discussion of metaphors for immigrants, see Amy L. Fairchild, *Science at the Borders: Immigrant Medical Inspection and the Shaping of the Modern Industrial Labor Force* (Baltimore, MD: Johns Hopkins University Press, 2003), 42–47.

21. Congressional Record 51—House, January 30, 1914, Debate on HR 6060 (a bill to regulate immigration), 2627.

22. Kenneth L. Roberts, "Slow Poison," *Saturday Evening Post,* June 26, 1920, 8–9, 54–58. Kenneth L. Roberts, "Canada Bars the Gates," *Saturday Evening Post,* August 12, 1922, 11; Lothrop Stoddard, "The New Realism of Science," *Saturday Evening Post,* September 6, 1921, 121.

23. French Strother, "The Immigration Peril," *World's Work,* October 1923, 634.

24. Cornelia James Cannon, "Selecting Citizens," *North American Review* (September 1923): 325.

25. Robert DeCourcy Ward, "The Immigration Problem," *Charities* 12 (February 1904): 141–42.

26. George Lydston, *The Diseases of Society and Degeneracy* (Philadelphia: J. B. Lippincott, 1908), 129, 132–33. See also "Making the Immigrant Unwelcome," *Literary Digest* 69, no. 5 (April 30, 1921): 34.

27. Irving Fisher, "Impending Problems of Eugenics," *Scientific Monthly* 13, no. 3 (September 1921): 227.

28. Letter from Williams to the Commissioner General, March 31, 1913, National Archives, Records of the Immigration and Naturalization Service, RG 85, entry 9, file 51490/19, p. 11; William Williams, "The New Immigration: Some Unfavorable Features and Possible Remedies," in *Proceedings of the National Conference of Charities and Correction 33* (Philadelphia: Press of Fred J. Heer, 1906), 7; William Williams, "Remarks on Immigration," unpublished address to the Senior Class of Princeton University, 1904, in William Williams Papers, New York Public Library, quoted in Alan M. Kraut, *Silent Travelers: Germs, Genes, and the "Immigrant Menace"* (Baltimore, MD: Johns Hopkins University Press, 1995), 68, 298 n. 57.

29. "Keep America 'White'!" *Current Opinion* 74 (April 1923): 399–401.

30. "The Invasion of the Unfit," *Medical Record* 82, no. 24 (December 14, 1912): 1080.

31. Alfred C. Reed, "The Medical Side of Immigration," *Popular Science Monthly,* April 1912, 384–90.

32. W. C. Billings, "The Prevention of Quarantinable Diseases on the Border and at Ports of Embarkation," *California State Journal of Medicine* 15, no. 5 (May 1917): 162.

33. Mary Douglas, *Natural Symbols: Explorations in Cosmology* (London: Routledge, 1973), xxxiii. Also quoted in John F. Kasson, *Rudeness and Civility: Manners in Nineteenth-Century Urban America* (New York: Hill & Wang, 1990), 124.

34. See Kasson, *Rudeness and Civility;* Karen Halttunen, *Confidence Men and Painted Women: A Study of Middle-Class Culture in America, 1830–1870* (New Haven, CT: Yale University Press, 1982).

35. Kasson, *Rudeness and Civility*, 44–45, 114–15, 118–23, 126.

36. Kathy Lee Peiss, *Hope in a Jar: The Making of America's Beauty Culture* (New York: Henry Holt, 1998), 48–49.

37. John Foster Carr, *Guide to the United States for the Immigrant Italian: A Nearly Literal Translation of the Italian Version* (New York: Doubleday, 1911), 71.

38. Quoted in Kasson, *Rudeness and Civility*, 167. Annie Randall White, *Polite Society at Home and Abroad: A Complete Compendium of Information upon All Topics Classified under the Head of Etiquette* (Chicago: L. P. Miller, 1891), 32. *The Manners That Win: Compiled from the Latest Authorities* (Minneapolis, MN: Buckeye, 1880), 353.

39. B. D. Pettingill, "The Sign-Language," *American Annals of the Deaf* 18 (January 1873): 4; "The Perversity of Deaf-Mutism," *American Annals of the Deaf* 18 (October 1873): 263; Sarah Harvey Porter, "The Suppression of Signs by Force," *American Annals of the Deaf* 39 (June 1894): 171. Porter made a similar observation in *Annals* 58 (1913): 284. See also Edward M. Gallaudet, "How Shall the Deaf Be Educated," *International Review* (December 1881), reprinted in Joseph C. Gordon, ed., *Education of Deaf Children: Evidence of Edward Miner Gallaudet and Alexander Graham Bell Presented to the Royal Commission of the United Kingdom* (Washington, DC: Volta Bureau, 1892), 101.

40. "Vulgarity in Signing," *Silent Educator* 1 (January 1890): 91.

41. B. Engelsman, "Deaf Mutes and Their Instruction," *Science* 16 (October 17, 1890): 220. See also Douglas C. Baynton, *Forbidden Signs: American Culture and the Campaign against Sign Language* (Chicago: University of Chicago Press, 1996), 52–54.

42. Sander L. Gilman, *Disease and Representation: Images of Illness from Madness to AIDS* (Ithaca, NY: Cornell University Press, 1988), 130–31.

43. Allan McLane Hamilton, *Types of Insanity: An Illustrated Guide in the Physical Diagnosis of Mental Disease* (New York: William Wood, 1883), 1.

44. Philip M Ferguson, *Abandoned to Their Fate: Social Policy and Practice toward Severely Retarded People in America, 1820–1920* (Philadelphia: Temple University Press, 1994), 5–6.

45. Pernick, *The Black Stork*, 60–61. See also Sander L Gilman, *Picturing Health and Illness: Images of Identity and Difference* (Baltimore, MD: Johns Hopkins University Press, 1995), 54.

46. Christopher Lasch, *Haven in a Heartless World: The Family Besieged* (New York: Norton, 1995); Tamara K. Hareven, "The Home and the Family in Historical Perspective," *Social Research* 58, no. 1 (Spring 1991): 253–85; Barbara Laslett, "The Family as a Public and Private Institution: An Historical Perspective," *Journal of Marriage and Family* 35, no. 3 (August 1973): 480–92; Halttunen, *Confidence Men and Painted Women*, 58–59.

47. Susan M. Schweik, *The Ugly Laws: Disability in Public* (New York: New York University Press, 2009), 291–96. See also Tobin Siebers, "What Can Disability Studies Learn from the Culture Wars?" *Cultural Critique* 55, no. 1 (2003): 198–201; Rosemarie Garland Thomson, *Extraordinary Bodies: Figuring Physical Disability in American Culture and Literature* (New York: Columbia University Press, 1997), 7.

48. Rosemarie Garland-Thomson, "Seeing the Disabled: Visual Rhetorics of Disability in Popular Photography," in *The New Disability History: American Perspectives*, ed. Paul K. Longmore and Lauri Umansky (New York: New York University Press, 2001), 338.

49. Frank B. Gilbreth and Lillian Moller Gilbreth, *Motion Study for the Handicapped* (London: G. Routledge, 1920), 27, 135–36.

50. James W. Trent, *Inventing the Feeble Mind: A History of Mental Retardation in the United States* (Berkeley and Los Angeles: University of California Press, 1994), 82, 142–44, 166–67;

David Wright, "Getting Out of the Asylum: Understanding the Confinement of the Insane in the Nineteenth Century," *Social History of Medicine* 10, no. 1 (April 1, 1997): 139, 146, 152–53. See also Gerald N. Grob, *Mental Illness and American Society, 1875–1940* (Princeton, NJ: Princeton University Press, 1983).

51. Paul Longmore, *Why I Burned My Book, and Other Essays on Disability* (Philadelphia: Temple University Press, 2003), 36.

52. Clayton, *Forgotten Prophet*, 220.

53. Albert Edward Wiggam, *The Fruit of the Family Tree* (Indianapolis: Bobbs-Merrill, 1924), 262, 271–73, 277–79. For references to immigrants as livestock, see "Putting Our Immigrants through the Sieve at Ellis Island," *Portland Oregonian*, September 28, 1913 ("herding the sheep from the goats"); John Watrous Knight, "The Working Man and Immigration," *Charities Review* 4 (1895): 365–66.

54. Edward Alsworth Ross, *The Old World in the New: The Significance of Past and Present Immigration to the American People* (New York: Century, 1914), 285–88.

55. Alfred C. Reed, "Going through Ellis Island," *Popular Science Monthly*, January 1913, 8–9.

56. Quoted in John T. E. Richardson, *Howard Andrew Knox: Pioneer of Intelligence Testing at Ellis Island* (New York: Columbia University Press, 2011), 40.

57. Letters, January 5 and January 10, 1906, National Archives, RG 90, entry 10, file 219. "Mrs. Thompson's Case to be Decided To-day," *New York Times*, January 9, 1906.

58. Letters, January 5 and January 10, 1906, National Archives, RG 90, entry 10, file 219.

59. Letters, January 5 and January 10, 1906, National Archives, RG 90, entry 10, file 219. *Use Ibid here instead?*

60. Carlisle P. Knight, "The Detection of the Mentally Defective among Immigrants," *Journal of the American Medical Association* 60 (1913): 106.

61. United States Public Health Service, *Regulations*, 16–19. The manual then went on to list diseases and disabilities that could be cause for exclusion—for instance, arthritis, asthma, bunions, deafness, deformities, flat feet, hernia, hysteria, poor eyesight, poor physical development, spinal curvature, and varicose veins. See also Victor Safford, *Immigration Problems: Personal Experiences of an Official* (New York: Dodd, Mead, 1925), 244–45.

62. Allan McLaughlin, "How Immigrants Are Inspected," *Popular Science Monthly*, February 1905, 359.

63. Kraut, *Silent Travelers*, 54–55.

64. E. H. Mullan, "The Medical Inspection of Immigrants at Ellis Island," *Medical Record* 84 (1913): 1185–87. See also E. H. Mullan, "Mental Examination of Immigrants: Administration and Line Inspection at Ellis Island," *Public Health Reports* 32, no. 20 (May 18, 1917): 734–35; J. G. Wilson, "Some Remarks Concerning Diagnosis by Inspection," *New York Medical Journal* 94 (1911): 94–96; Richardson, *Howard Andrew Knox*, 42; Amy L. Fairchild, *Science at the Borders: Immigration Medical Inspection and the Shaping of the Modern Industrial Labor Force* (Baltimore, MD: Johns Hopkins University Press, 2003), 83–92; William C. Van Vleck, *The Administrative Control of Aliens: A Study in Administrative Law and Procedure* (New York: Commonwealth Fund, 1932), 28.

65. Quoted in Elizabeth Yew, "Medical Inspection of Immigrants at Ellis Island, 1891–1924," *Bulletin of the New York Academy of Medicine* 56, no. 5 (June 1980): 497–98.

66. Safford, *Immigration Problems*, 244–46.

67. Frederic Haskin, *The Immigrant: An Asset and a Liability* (New York: Fleming H. Revell, 1913), 76. Quoted in Anne-Emmanuelle Birn, "Six Seconds per Eyelid: The Medical Inspection of Immigrants at Ellis Island, 1892–1914," *Dynamis* 17 (1997): 290.

68. "Putting Our Immigrants through the Sieve," *Portland Oregonian*, September 28, 1913.

69. *Physical Examination of Immigrants: Hearings before the Committee on Immigration and Naturalization*, H.R. 66th Cong., 3d Sess., 11 (January 11, 1921). Amy Fairchild, in her important study of the inspection process, explores this topic and argues that while this "made possible a search for only a limited range of conditions visibly affecting the ability of the immigrant to work," it was significant by analogy to the assembly line as an initiation into a society based on industrial discipline; Fairchild, *Science at the Borders*, 86.

70. Deborah L. Rhode, *The Beauty Bias: The Injustice of Appearance in Life and Law* (Oxford: Oxford University Press, 2010). Daniel S. Hamermesh, *Beauty Pays: Why Attractive People Are More Successful* (Princeton, NJ: Princeton University Press, 2011). Nicholas Bakalar, "Ugly Children May Get Parental Short Shrift," *New York Times*, May 3, 2005.

71. National Archives, Records of the INS, RG 85, entry 9, file 51497/4. National Archives, Records of the INS, RG 85, Accession 60A600, file 52880/113. (Both Fruman's and Mydland's stories are told in chapter 1.)

72. Case of Angelo Berardis of Italy, 1906, National Archives, Records of the INS, RG 85, entry 7, file 50014/3. Sprios Tzoupis of Turkey, 1910, National Archives, Records of the INS, RG 85, Accession 60A600, file 52880/112. Janos Matiscak of Hungary, 1910, National Archives, Records of the INS, RG 85, Accession 60A600, file 52880/136. Jacob Dweck of Turkey, 1910, National Archives, Records of the INS, RG 85, Accession 60A600, file 52880/127.

73. "Conference to Consider Medical Examination of Immigrants—Minutes," February 8, 1907, National Archives, Record Group 85, Records of the INS, entry 9, file no. 51490/19.

74. National Archives, Record Group 85, Records of the INS, entry 7, file no. 48,462, March 6, 1905.

75. United States v. Petkos, Circuit Court of Appeals, 1st Cir., June 24, 1914, No. 1038, United States Courts: Circuit Court of Appeals, *Reports Containing the Cases Determined in All the Circuits from the Organization of the Courts* (West Publishing Company, 1915), 274; Ex Parte Petkos, District Court, D. Massachusetts, April 18, 1913, No. 736, Robert Desty, James Wells Goodwin, and Peyton Boyle, *The Federal Reporter: With Key-Number Annotations* (St. Paul, MN: West, 1914), 275; see also United States Bureau of Immigration, *Annual Report* (Washington, DC: U.S. Government Printing Office, 1914), 216.

76. National Archives, Record Group 85, Records of the INS, entry 7, file no. 49951–1 (the immigrant's name is rendered as Abram Hofmann and Abram Hofman by immigration officials, Abraham Hoffman by his attorney). Spinal curvature was a common reason for rejection. A longitudinal study at the University of Iowa concluded that persons with late-onset scoliosis (occurring during puberty) "are productive and functional at a high level at 50-year follow-up" and experienced "little physical impairment," with "cosmetic concerns" being the only significant problem. This finding contradicts the common perception among physicians and the general public that this is a serious and debilitating condition; see S. L. Weinstein, L. A. Dolan, K. F. Spratt, K. K. Peterson, M. J. Spoonamore, and I. V. Ponseti, "Health and Function of Patients with Untreated Idiopathic Scoliosis: A 50-Year Natural History Study," *Journal of the American Medical Association* 289 (February 5, 2003): 559–68.

77. National Archives, Record Group 85, Records of the INS, entry 7, file no. 499144–3.

78. Quoted in Ian Robert Dowbiggin, *Keeping America Sane: Psychiatry and Eugenics in the United States and Canada, 1880–1940* (Ithaca, NY: Cornell University Press, 1997), 203–4.

79. Quoted in Kraut, *Silent Travelers*, 71.

80. Richardson, *Howard Andrew Knox*, 72.

81. "Current Comment," *Journal of the American Medical Association* 60, no. 2 (January 11, 1913): 133–35.

82. "Feeblemindedness and Immigration," *Journal of the American Medical Association* 60, no. 2 (January 11, 1913): 129.

83. Quoted in Desmond S. King, *In the Name of Liberalism: Illiberal Social Policy in the USA and Britain* (Oxford: Oxford University Press, 2004), 112.

84. Letter from F. P. Sargent, Commissioner General, Bureau of Immigration, to Commissioner at Ellis Island [Robert Watchorn], September 13, 1905, National Archives, Record Group 90, Records of the INS, entry 10, box 36, file 219.

85. United States, Bureau of Immigration, *Annual Report of the Commissioner-General of Immigration to the Secretary of the Treasury for the Fiscal Year Ended June 30, 1911* (Washington, DC: Government Printing Office, 1912), 147. See also Alfred C. Reed, "The Medical Side of Immigration," *Popular Science Monthly*, April 1912, 384–90.

86. Carlisle P. Knight, "The Detection of the Mentally Defective among Immigrants," *Journal of the American Medical Association* 60 (1913): 106–7.

87. Mullan, "Mental Examination of Immigrants," 735, 738.

88. Fiorella H. La Guardia, *The Making of an Insurgent* (New York: J. B. Lippincott, 1948), 65.

89. Richardson, *Howard Andrew Knox*, 70. See also Fairchild, *Science at the Borders*, 102–3. Richardson cited Leila Zenderland, *Measuring Minds: Henry Herbert Goddard and the Origins of American Intelligence Testing* (Cambridge: Cambridge University Press, 1998), 266.

90. Zenderland, *Measuring Minds*, 271–77. Howard Knox, *Manual of the Mental Examination of Aliens* (Washington, DC: Government Printing Office, 1918). See also Mullan, "Mental Examination of Immigrants," 746.

91. Howard A. Knox, "A Diagnostic Study of the Face," *New York Medical Journal* (June 14, 1913): 1225–31.

92. Ibid., 1227–28.

93. National Archives, Records of the INS, RG 85, entry 7, file no. 48,599/4.

94. National Archives, Records of the INS, RG 85, Accession 60A600, file 52,242–11.

95. National Archives, Records of the INS, RG 85, Accession 60A600, file 53,248–18 (April 28, 1911).

96. National Archives, Records of the INS, RG 85, Accession 60A600, file 53452–952.

97. National Archives, Records of District Courts of the United States, Records of the U.S. Circuit Court for the Southern District, RG 21, entry 34.

98. Letter from W. W. Husband, Commissioner General, Bureau of Immigration, to H. S. Cumming, Surgeon General, United States Public Health Service, September 27, 1922; and reply from Cumming to Husband, September 29, 1922; National Archives, RG 90, entry 10, file 219.

99. National Archives, Record Group 85, Records of the INS, Accession No. 60A600, file 51,806–16.

Conclusion

1. United States Government, *Abstracts of Reports of the Immigration Commission*, vol. 1 (Washington, DC: Government Printing Office, 1911), 35.

2. *Annual Report of the Commissioner-General of Immigration* (Washington, DC: Government Printing Office, 1912), 123.

3. Figures up until 1910 can be found in United States Government, Abstracts of Reports of the Immigration Commission, vol. 1, 110. After that, statistics are included in the annual reports of the commissioner general. The 1912 commissioner general's report states that while 3,055 were

rejected as mentally or physically defective, 12,004 were rejected as "likely to become a public charge," and of those "a considerable portion . . . were excluded on the additional ground of being mentally or physically defective. . . . Where the exclusion occurs on both grounds, it is not an easy matter properly to classify the cases in the statistical reports; and the figures representing those 'LPC' and those 'mentally and physically defective' should be considered together." Annual Report of the Commissioner-General of Immigration (1912), 125. The claim that wartime inspection was more thorough and therefore more effective can be found in, among other sources, James Davis, Selective Immigration (St. Paul, MN: Scott Mitchell, 1925), 47. Davis was Secretary of Labor from 1921 to 1930.

4. *U.S. Statutes*, vol. 26 (1891), 1086; *U.S. Statutes*, vol. 27 (1893), 569; *U.S. Statutes*, vol. 34 (1907), 901–902. See also Joseph H. Senner, "How We Restrict Immigration," *North American Review* 158 (1894): 498.

5. U.S. Immigration Service, *Annual Report of the Superintendent of Immigration* (Washington, DC: Government Printing Office, 1894), 12–13. See also Edith Abbott, *Immigration: Select Documents and Case Records* (Chicago: University of Chicago Press, 1924), 71; U.S. Bureau of Immigration, *Annual Report of the Commissioner-General of Immigration* (Washington, DC: Government Printing Office, 1907), 62, 83; Max J. Kohler, "Restriction of Immigration—Discussion," *American Economic Review* 2, no. 1 (March 1912), 75; Amy L. Fairchild, *Science at the Borders: Immigrant Medical Inspection and the Shaping of the Modern Industrial Labor Force* (Baltimore, MD: Johns Hopkins University Press, 2003), 56–63; Penny Richards, "Points of Entry: Disability and the Historical Geography of Immigration," *Disability Studies Quarterly* 24, no. 3 (Summer 2004), http://dsq-sds.org/article/view/505/682; Jenna Weissman Joselit, "The Perceptions and Reality of Immigrant Health Conditions, 1840–1920," in *U.S. Immigration Policy and the National Interest: Staff Report of the Select Commission on Immigration and Refugee Policy: Supplement to the Final Report and Recommendations, Appendix A* (Washington, DC: The Commission, 1981), 209–10.

6. Henry C. Hansbrough, "Why Immigration Should Not Be Suspended," *North American Review* 156 (1893): 224.

7. Senner, "How We Restrict Immigration," 498.

8. Prescott Hall, "Immigration and the Educational Test," *North American Review* 165 (1897): 400.

9. Letter to F. P. Sargent, January 25, 1904, and reply, January 27, 1904, New York Public Library, William Williams Papers, 1902–1943, Box 1, No. 20897.

10. "Immigration Record Will Be Broken This Year; Yet Commissioner Watchorn Sees No Grounds for Alarm," *New York Times*, March 11, 1906. In some cases, however, shipping companies were known to add a surcharge to the fare of disabled immigrants to defray the costs of returning them if rejected. This was noted in the case of Leopold Perelman, who apparently had paid extra on account of his deafness; National Archives, Records of the Immigration and Naturalization Service, RG 85, entry 7, file 49731/2 (he arrived at the Port of Boston on December 4, 1905).

11. James Davenport Whelpley, "The Open Door for Immigrants," *Harper's Weekly*, April 14, 1906, 517–19.

12. *Abstract of Reports of the Immigration Commission*, vol. 1 (Washington, DC: Government Printing Office, 1911), 26, 170.

13. "People We Bar Out: Big Increase in Deportation of the Immigrant Classes," *Chicago Inter Ocean*, February 11, 1900.

14. Fairchild, *Science at the Borders*, 68.

15. Jennifer Blakeman, "The Exclusion of Mentally Ill Aliens Who May Pose a Danger to Others: Where Does the Real Threat Lie?" *Inter-American Law Review* 31, no. 2 (2000): 295–97.

16. Fairchild, *Science at the Borders*, 259–64, 270–71.

17. Blakeman, "The Exclusion of Mentally Ill Aliens," 299–301.

18. Valerie Hauch, "Disabled Woman Denied Entry to U.S. After Agent Cites Supposedly Private Medical Details," *Toronto Star*, November 28, 2013, http://www.thestar.com/news/gta/2013/11/28/disabled_woman_denied_entry_to_us_after_agent_cites_supposedly_private_medical_details.html.

19. Sarah Bridge, "Canadians with Mental Illnesses Denied U.S. Entry," *Toronto Star*, September 9, 2011, http://www.cbc.ca/news/canada/canadians-with-mental-illnesses-denied-u-s-entry-1.1034903. See also: Isabel Teotonio, "Canadian Woman Denied Entry to U.S. Because of Suicide Attempt," *Toronto Star*, January 29, 2011, http://www.thestar.com/news/gta/2011/01/29/canadian_woman_denied_entry_to_us_because_of_suicide_attempt.html.

20. Mark C. Weber, "Opening the Golden Door: Disability and the Law of Immigration," *Journal of Gender, Race and Justice* 8 (Spring 2004): 162–63, 168 n. 76.

21. John F. Stanton, "The Immigration Laws from a Disability Perspective: Where We Were, Where We Are, Where We Should Be," *Georgetown Immigration Law Journal* 10 (Spring 1996): 441–65. When Stanton's article went to press, the family was considering an appeal. Stanton later lost track of the case and does not know its outcome (personal communication, October 25, 2005.).

22. Kathrin S. Mautino, "Immigration and Physical Disability," *Journal of Immigrant Health* 4, no. 2 (April 2002): 60–61.

23. Ellsworth Eliot, Jr., M.D., "Immigration," in Madison Grant and Charles Stewart Davison, *The Alien in Our Midst, or Selling Our Birthright for a Mess of Pottage* (New York: Galton, 1930), 101.

24. Quoted in Joseph P. Shapiro, *No Pity: People with Disabilities Forging a New Civil Rights Movement* (New York: Times Books, 1993), 165.

Index

Page numbers in italics refer to illustrations.